Coercion, Survival, and War

Coercion, Survival, and War

WHY WEAK STATES RESIST THE UNITED STATES

Phil Haun

Stanford Security Studies
An Imprint of Stanford University Press
Stanford, California

Stanford University Press
Stanford, California

Printed in the United States of America on acid-free, archival-quality paper

Library of Congress Cataloging-in-Publication Data

Haun, Phil M., author.
    Coercion, survival, and war : why weak states resist the United States / Phil Haun.
        pages cm
    Includes bibliographical references and index.
    ISBN 978-0-8047-9283-7 (cloth : alk. paper)
        1. Asymmetric warfare—United States—Case studies. 2. Asymmetric warfare—
    Case studies. 3. United States—Military policy—Case studies. 4. United States—
    Foreign relations—Case studies. I. Title.
    U163.H39 2015
    355.4′2—dc23

                                                                    2014036166

ISBN 978-0-8047-9507-4 (electronic)

Typeset by Thompson Type in 10/14 Minion

To Bonnie for filling my life with such joy

CONTENTS

TABLES, FIGURES, AND MAPS

Tables

Figures

Maps

ACKNOWLEDGMENTS

My journey to answer why the United States fails to coerce weak states commenced in the spring of 1999 when, as an A-10 pilot, I dodged missiles attempting to locate, identify, and destroy Serbian armor. No one up the chain of command could explain how NATO's air strikes would convince Slobodan Milosevic to cede Kosovo. Subsequently, at the School of Advanced Air Power Studies, I studied with Stephen Chiabotti, Forrest Morgan, David Mets, Hal Winton, Rich Muller, Ev Dolman, Jim Corum, Richard Andres, and Tom Ehrhard, who helped me understand how military force could be applied to achieve political outcomes. Many years later at the Security Studies Program at MIT I discovered a group of intellectuals led by Barry Posen, who refused to cut corners in the quest for knowledge or employ methodological gimmicks in the search for truth. I am indebted to Owen Cote, Ken Oye, Robert Art, Roger Petersen, Richard Samuels, Stephen Van Evera, Harvey Sapolsky, Taylor Fravel, Fotini Christia, David Singer, Gabe Lenz, Cindy Williams, Jim Walsh, and Vipin Narang and to my cohort and colleagues Josh Shifrinson, Tara Maller, Austin Long, Caitlin Talmadge, Paul Staniland, Keren Fraiman, Brendan Green, Jon Lindsay, Andrew Radin, Peter Krause, Dan Altman, Will Norris, Nathan Black, Kentaro Maeda, Rachel Wellhausen, Stephanie Kaplan, Sameer Lalwani, Gautam Mukunda, Ben Friedman, Kelly Grieco, Chris Clary, Miranda Priebe, and Jill Hazelton. I thank Barbara Geddes, Ken Schultz, and Todd Sechser for providing me access to their databases and Andrew Bennett, Bob Pape, and Robert Powell for their advice. I especially thank an unnamed reviewer for his or her ruthless feedback. I thank my colleagues at the U.S.

Naval War College, especially John Maurer, Colin Jackson, Josh Rovner, Brad Lee, Wick Murray, Scott Douglas, Tim Schultz, Tim Hoyt, Sally Paine, Charlie Edel, Tom Mahnken, David Kaiser, Karl Walling, Dex Wilson, David Brown, Bob Flynn, Mike Vlahos, Toshi Yoshihara, and Nick Saradakis. I thank Fairchild Republic for making an aircraft to survive missile strikes, God for watching over me, and my family for making it all worthwhile. Finally this book is dedicated to Bonnie, who has been with me through it all. Thank you for your talent, patience, and sacrifice to bring this book into reality.

Coercion, Survival, and War

1 INTRODUCTION

ON AUGUST 2, 1990, ARMORED DIVISIONS of the Iraqi Republican Guard rolled into Kuwait City. The United States responded immediately, convening the UN Security Council, condemning the aggression, and demanding an immediate and unconditional withdrawal. Within days, the Security Council imposed comprehensive sanctions to compel Iraq to reverse its actions. Meanwhile, the U.S. military deployed thousands of aircraft, ships, and troops to the Middle East to deter further aggression. Though economic sanctions soon forced the Iraqi government to begin rationing, its dictator, Saddam Hussein, remained obstinate. In October, President George H. W. Bush ordered a doubling of U.S. deployed forces with the objectives of liberating Kuwait and destroying the Iraqi Army. To garner international and domestic support for war, the administration pressed the Security Council to authorize a coalition of thirty-four nations to use all means necessary to free Kuwait should the Iraqi Army not withdraw by January 15, 1991. In the final days leading up to the deadline, however, the White House feared not that coercion would fail but that it might succeed. Bush had already paid the diplomatic, political, and deployment costs of preparing for war, and his worst-case scenario now had Saddam capitulating. Although such an outcome would have rescued Kuwait, it would also have spared the Republican Guard, guaranteeing Saddam would remain in power and threaten his neighbors in the future.

Fortunately for Bush, coercive diplomacy failed when the January 15 deadline expired and the U.S.-led coalition launched a massive air campaign. Coalition fighters quickly secured air superiority over Kuwait and southern Iraq, while its bombers and cruise missiles simultaneously attacked leadership

1

targets in Baghdad in an effort to paralyze Saddam's ability to command and control his military. Undeterred, Saddam ordered Scud missiles to be launched at Israel and Saudi Arabia and, in late January, directed forward elements of his army to advance into Saudi Arabia in an attempt to jump-start his "mother of all battles." By early February, coalition air power shifted its effort toward attriting the Iraqi Army in preparation for the land invasion. Saddam's strategy had failed both to draw Israel into the conflict and to bait the coalition into an early ground war. His hope of victory faded as the air strikes against Iraqi fielded forces intensified. By mid-February, Iraqi officials broached the possibility of a cease-fire, and Soviet diplomats initiated discussions on the terms of a peace proposal. Fearful that Saddam would now agree to the UN resolutions and deny it the opportunity to destroy the Republican Guard, the United States accelerated the start date for the ground offensive. On the eve of the invasion, Saddam accepted the Soviet-brokered deal, which would have suspended all sanctions against Iraq on the withdrawal of its army from Kuwait, a process for which Iraq would be granted twenty-one days. Bush rejected these conditional terms and instead issued a forty-eight-hour ultimatum for withdrawal. To meet such a deadline would have required the Iraqi Army to abandon its heavy weaponry and prepared defenses. Though willing to withdraw from Kuwait in an orderly fashion, Saddam would not accept the humiliation of a hasty retreat, which would have left Iraq's southern border exposed and his regime vulnerable to domestic unrest. Coercion failed as Saddam chose to face the ground war with a slim chance of victory rather than concede, which would have threatened the survival of the Iraqi state and his regime.

This book is about why the United States chooses to coerce weak states and, as with Iraq in 1991, why coercion so often fails. Since World War II, the United States has been the great power that has most frequently initiated crises with much weaker states.[1] In these asymmetric interstate crises, U.S. leaders have more often than not chosen coercion, which has been met by resistance in half the cases. And, of these unsuccessful efforts, half again have been resolved only by war. Given the overwhelming military advantage enjoyed by the United States, why do weak states so often resist, and why does coercion fail?

Coercion threatens force or employs limited force to convince a target to comply with a challenger's demands.[2] In a conflict against a weak state, the United States, with its disproportionate advantage in power, has the luxury of choice, to decide whether to accommodate or to escalate the disagreement into a crisis by threatening military force. When choosing force, the United States has the options of coercion or brute force war. Because coercion requires the cooperation of the targeted state, it is a strategy likely to achieve only moderate foreign policy goals but with the advantage of avoiding the full costs of war. This is an attractive option for U.S. decision makers, particularly when conflicts are over issues nonvital to national security. Brute force strategies, by contrast, do not require the cooperation of the enemy to succeed and may thus realize more ambitious objectives, such as seizing territory or imposing regime change. Such actions, however, incur a higher price in terms of blood and treasure. The United States therefore usually prefers coercion, even with its more moderate goals and diminished chance of success.

International relations scholars offer numerous reasons for why coercion fails. Much of the literature focuses on psychological, cognitive, or organizational biases and the resultant nonrational decision making that leads to war.[3] Others offer structural causes of war based on uncertainty or commitment problems inherent to the anarchic international system. These arguments, although useful for explaining why crises arise or why violence erupts, fall short of explaining the final outcome for most cases of asymmetric interstate coercion.

In theory, coercion should always succeed, as the United Sates should choose a coercive strategy only when it expects the targeted state will acquiesce. In reality, however, coercion often fails, as policy makers mismatch demands and threats. In some cases, the demands are so severe that, if conceded, they jeopardize the survival of the much weaker state or its regime. When this is the case, there is not likely to be any credible threat that will convince a target to agree to its own demise, so long that it retains the means to resist.[4]

A state's survival requires that it retain sovereignty over its domestic and international affairs. Domestic sovereignty is threatened when a powerful challenger demands that a targeted state concede homeland territory. In Bosnia from 1992 to 1995, the Bosnian Serbs consistently resisted any peace deal that required them to concede territory. Though the Republika Srbska had not been recognized as a state, it acted as one. Only after it had already lost its western lands to the Croatian Muslim ground offensive backed by NATO air power did it finally agree to concede this territory, which it no longer possessed, to retain the ground it still held. Rationally, a powerful challenger should understand that a target will resist in such cases as its survival depends on it protecting its homeland.

Along with homeland territory, a state's survival requires that it maintain its international sovereignty over its foreign policy decision making. A state will thus reject demands for regime change, as this forfeits such autonomy. In March 2003, for example, Saddam rejected President George W. Bush's demand that he leave Iraq only days before the U.S. invasion. Although leaders will refuse to give up a state's autonomy, if subjected to sufficient coercive pressure they may be constrained in their foreign policy choices. For example, after ordering an invasion of Jordan in September 1970, Syrian leaders reversed this decision within days. An effective defense by the Jordanians, combined with the threat of the Israeli Air Force backed by U.S. forces in the Mediterranean, deterred the Syrians from committing their air force to the battle. As a result, this compelled the Syrians to withdraw their ground forces from Jordan.[5] Coercion succeeded, in part, because Syria's decision to withhold its air power did not threaten its survival, whereas having its air force destroyed by the Israelis may well have done so.

In addition to state survival, a ruling regime must also be wary of the potential threat from domestic opposition. Concessions signal that the regime is weak, a revelation that may trigger a coup or insurgency. Saddam resisted Bush's forty-eight-hour ultimatum in 1991 out of concern over the audience costs associated with such a public foreign policy failure. The Shia and Kurdish revolt after the Gulf War suggests that these fears were well founded.

Survival concerns are particularly relevant in asymmetric interstate crises, given the inherent power disparities between great powers and weak states. The tremendous military advantage enjoyed by a powerful challenger enhances its expected outcome in a war against a weak opponent. Ceteris paribus, to prefer coercion over brute force requires the great power to levy heavy demands on its target, which, in turn, increases the likelihood concessions will threaten the survival of the weak state and its regime. This being the case, however, the target is likely to resist and coercion fail.

The resistance of weak states introduces a related question. Why would an American statesman choose a coercive strategy likely to fail? If the demands are high, such that the weak state is not likely to be coerced, why doesn't the United States simply take its objectives by force? Again, the international relations literature refers us to the previously mentioned list of rational and nonrational explanations for war. In addition, two incentives exist for a powerful challenger to intentionally choose a coercive strategy it expects to fail. First, coercion may be but an interim stage of a brute force strategy. It takes time to mobilize and deploy military forces to invade and occupy a country. If the costs of coercion are low, and there is some chance that the weak state might concede, then it may make sense to issue coercive demands while preparing for war.

Second, modern states are expected to seek resolution to conflicts through negotiations or intervention by international institutions before resorting to violence. U.S. leaders thus have an incentive to work through the UN Security Council to avoid the diplomatic and political costs of abrogating this norm. To justify war, the United States may therefore make the pretense of seeking a settlement when it, in fact, prefers that the weak state reject its efforts. This proved to be the case in the lead-up to the 2003 invasion of Iraq. President Bush spoke before the UN General Assembly on September 12, 2002, demanding that Iraq disclose and abandon its weapons of mass destruction (WMD). Before the Security Council could even pass a resolution, however, he began making the case that Saddam could not be trusted.[6] Bush counted on coercive diplomacy to fail and thus provide justification for the war he had already decided on.

U.S. COERCION SINCE WORLD WAR II

Since the Second World War, the United States has been involved in half of the asymmetric interstate crises pitting great powers against much weaker states.[7] Table 1.1 lists the cases in which the United States threatened or employed military force. In twenty-three of these thirty conflicts, U.S. leaders opted for coercion over brute force. Chapters 4 through 6 provide in-depth assessments of six of these crises with Iraq, Serbia, and Libya, respectively. Appendix A summarizes the remaining seventeen coercive cases. The summaries include the rationale for the coding provided in Table 1.1 for foreign policy, coercion, and coercive diplomacy success or failure. For a case to be coded as a success, the United States must have concluded the crisis having attained its core ex ante objectives. Overall, the United States achieved these foreign policy goals two-thirds of the time (twenty of thirty).

Coercion is coded as a failure when it did not achieve U.S. core foreign policy objectives or when the United States abandoned coercion in favor of a brute force strategy. This latter reason has increasingly caused coercion failure. It occurred in five out of the twelve coercion cases that have taken place since the Cold War: in 1991 in the Gulf War, in 1998 in the Desert Fox air

Table 1.1. U.S. Asymmetric Interstate Crises: 1950–2011.

Year	Target state	Crisis name	U.S. strategy	Foreign policy outcome	Coercion Outcome	Coercive diplomacy outcome
1950	North Korea	Korean War	Brute force	Success		
1961	Cuba	Bay of Pigs	Brute force	Failure		
1964	North Vietnam	Gulf of Tonkin	Coercion	Failure	Failure	Failure
1965	North Vietnam	Pleiku/Rolling Thunder	Coercion	Failure	Failure	
1968	North Vietnam	Tet Offensive	Coercion	Failure	Failure	
1970	Syria	Black September	Coercion	Success	Success	Success
1972	North Vietnam	Linebacker I	Coercion	Success	Success	
1972	North Vietnam	Linebacker II	Coercion	Success	Success	
1975	Cambodia	Mayaguez	Coercion	Success	Success	Failure
1976	North Korea	Poplar Tree	Coercion	Success	Success	Success
1979	Iran	U.S. hostages	Brute force	Failure		
1981	Libya	Gulf of Sidra	Coercion	Failure	Failure	Failure
1983	Grenada	Grenada invasion	Brute force	Success		
1986	Libya	El Dorado Canyon	Coercion	Failure	Failure	Failure
1988	Nicaragua	Contras	Coercion	Failure	Failure	Failure
1989	Panama	Panama invasion	Brute force	Success		
1991	Iraq	Gulf War	Coercion/brute force	Success	Failure	Failure
1992	Serbia	Bosnia Civil War	Coercion	Success	Success	Failure
1992	Iraq	Iraq no-fly zone	Brute force	Success		
1993	North Korea	Nuclear crisis	Coercion	Success	Success	Success
1994	Haiti	Military regime	Coercion	Success	Success	Success
1994	Iraq	Kuwaiti border	Coercion	Success	Success	Success
1996	Iraq	Desert Strike	Brute force	Success		
1997	Iraq	Desert Fox	Coercion/brute force	Failure	Failure	Failure
1998	Afghanistan	Embassy bombing	Coercion	Failure	Failure	Failure
1999	Serbia	Kosovo	Coercion	Success	Success	Failure
2001	Afghanistan	Afghanistan War	Coercion/brute force	Success	Failure	Failure
2003	Iraq	Regime change	Coercion/brute force	Success	Failure	Failure
2003	Libya	WMD	Coercion	Success	Success	Success
2011	Libya	Arab Spring	Coercion/brute force	Success	Failure	Failure

SOURCE: International Development and Conflict Management's International Crisis Behavior (ICB) database and Kenneth Schultz and Jeffrey Lewis's Coercive Diplomacy Database.

NOTE: Asymmetric interstate crises identified from an examination of the Center for International Development and Conflict Management's International Crisis Behavior (ICB) database compared to Kenneth Schultz and Jeffrey Lewis's Coercive Diplomacy Database, with which they expand the ICB database into dyadic cases.

strikes, in 2001 in the Afghanistan War, in 2003 in the Iraq War, and in 2011 in the Libyan Arab Spring.

Coercive diplomacy is a more restrictive form of coercion, for which violence is only threatened. Coercive diplomacy is not applicable for many cases of interwar coercion, in which force is already being employed. Where coercive diplomacy has been attempted, it has failed in two-thirds of the cases (thirteen of nineteen).[8] This lower success rate is understandable because, for coercion to succeed, a powerful challenger must at times first demonstrate resolve by employing limited force. Other times, coercive diplomacy fails because demands are initially too high and the threats too low. In such cases, coercive diplomacy fails, but coercion eventually succeeds when, over time, the powerful challenger adjusts its threats and demands until the weak target acquiesces.

Because war, coercion, and brute force all involve a state's employment of violence, a clarifying comment is warranted. In theory, war is a state's use of violence to achieve its political objectives, but, in practice, the amount of violence required to call a crisis a war is set arbitrarily.[9] For instance, El Dorado Canyon, the coercive strategy that involved a single U.S. air raid against Libya, is not classified as a war even though blood was shed on both sides. By contrast, the eleven weeks of air strikes against Serbia over Kosovo are called a war, even though no NATO personnel were killed in combat. Whereas coercive strategies may or may not reach the requisite level of violence of being labeled as war, brute force strategies, where objectives are seized by force, such as in Korea, Vietnam, Afghanistan, and Iraq, are called wars. Because this book is more concerned with the U.S. decision making in choosing between coercive and brute force strategies and less concerned as to whether a particular crisis qualifies as a war, in an effort to limit confusion, the term *war* in this book will refer only to brute force strategies.

This book makes two primary contributions to our understanding of international politics and U.S. foreign policy. First, it introduces a theory for coercion specific to asymmetric interstate conflicts. A simple but powerful game theoretic model captures the unique and key characteristics of the interaction between a powerful challenger and a much weaker opponent. This model incorporates the alternative strategies available to the United States, that is, those of accommodation and brute force. Also, the coercion range is developed to demonstrate the set of potential demands for which the United States prefers a negotiated settlement to war and the weak target prefers concession to resistance.

This book further demonstrates that the survival concerns of weak states and their leaders provide a better explanation for coercion failure than the rational explanation for war based on commitment problems. The survival argument proposes that a weak state will resist the demands of a great power because concession would result in the loss of the state's and/or the regime's sovereignty. By contrast, the commitment explanation claims that a weak state's resistance is based less on concerns over the outcome of the present crisis and more on the impact a concession would have on future crises.

Second, this book presents seven historical cases drawn from recent U.S. experience with Iraq, Serbia, and Libya. Each case traces the decisions made by the United States and its adversaries that led ultimately to coercion success or failure. Together, these cases validate the asymmetric coercion model for analyzing asymmetric conflicts and demonstrate the significant role survival plays in the decision making of a weak state and its leaders.

This work should be of interest to foreign policy practitioners. It identifies clear limits to the effectiveness of coercive strategies. Demands, which threaten the survival of not only the weak state but also its regime, are most likely to be resisted. It also highlights that, at times, coercive diplomacy must first fail before coercion can succeed. Weak state leaders often face internal threats, prompting them to initially resist demands and provoke a limited U.S. attack to exhibit their resolve to those who might otherwise remove them from power.

This book also offers policy recommendations that increase the likelihood of coercion success for the United States in future asymmetric conflicts. Policy makers should be concerned not only with matching threats to demands but also with the manner in which their coercive strategies are implemented. The timing and delivery of coercive signals have a direct impact on the audience costs a weak state's leader will suffer for making concessions. In addition, U.S. leaders should continue to take steps to minimize commitment problems by working through international institutions, building coalitions, engaging third parties in the negotiations, and/or by implementing agreements incrementally.

OVERVIEW

This book proceeds as follows: Chapter 2 develops a theory of asymmetric coercion that concludes that the United States, as a powerful challenger, should coerce only when its weak target state is likely to concede and only then if this expected outcome is better than that of either accommodation or brute force war. The equilibrium conditions for this model suggest that coercion should then succeed, and a coercion range is developed for the set of demands to which both the challenger and the target will agree. In the real world, of course, coercion often fails, and the latter part of Chapter 2 introduces the conventional rational and nonrational explanations. Chapter 3 introduces an alternative explanation for coercion failure based on the target's survival concerns. The chapter concludes by considering two incentives for the United States to knowingly choose coercive strategies likely to fail.

Chapters 4 through 6 expand this analysis with a series of cases of U.S. asymmetric interstate coercion. Chapter 4 examines two crises between the United States and Iraq. In the 1991 Gulf War and the 2003 Iraq War, the United States adopted coercive strategies that threatened the survival of Saddam's regime and the Iraqi state. These crises test the limits for what coercion can achieve and examine the trade-offs between coercive and brute force strategies. In both crises, U.S. administrations chose coercive strategies they did not intend to have succeed in order to then implement the brute force strategies they preferred. These two crises thus provides insight into the key questions addressed by this book as to why the United States so often chooses coercion, why coercion so often fails as weak states resist, and, knowing this, why U.S decision makers might still prefer coercive strategies.

Chapter 5 examines two crises of the United States against the Bosnian Serbs and Serbia in 1992 and 1999, respectively. In both cases, coercive diplomacy failed, but coercion ultimately succeeded. The first crisis arose over actions in the Bosnian Civil War from 1992 to 1995, whereby the United States finally coerced the Bosnian Serbs into accepting a peace agreement but could never compel them to give up territory until it had already been taken by force. The second crisis arose over Serbia's treatment of Kosovar Albanians and concluded when Serbia's President Slobodan Milosevic finally conceded Kosovo after a seventy-eight-day NATO air campaign. He conceded, however, only when the expected economic costs to Serbia from the air campaign outweighed the political value of maintaining control of Kosovo. This crisis is a rigorous test of the survival hypothesis as Serbia eventually conceded homeland territory although it still retained the means to resist.

Chapter 6 considers three cases between the United States and Libya from 1981 until 2003. Libya's support of international terrorism triggered a crisis for the United States, culminating in the El Dorado Canyon air raid in April

1986. Coercion ultimately failed because of the mismatch between demands and threats as an isolated Reagan administration could not maintain the credible threat of force. The second case commenced with the bombing of Pan Am Flight 103 over Lockerbie, Scotland, in December 1988. It concluded with the extradition of two Libyan officials in 1999 to stand trial at The Hague. Another crisis commenced in September of 2002 when British Prime Minister Tony Blair, with the backing of George W. Bush, made overtures to Muammar Qaddafi to resume negotiations over Libya's WMD. The case concluded in December 2003 when Qaddafi abandoned Libya's nuclear, biological, and chemical ambitions.

2 A THEORY OF ASYMMETRIC
INTERSTATE COERCION

ON OCTOBER 4, 2003, OFFICIALS INTERDICTED a shipment bound for Libya containing components for centrifuges to enrich uranium. Caught red handed with irrefutable evidence of a nuclear weapons program, Colonel Muammar Qaddafi finally succumbed to U.S. coercive demands to abandon Libya's weapons of mass destruction (WMD) programs. This incident is just one example of the United States confronting a much weaker state. In such asymmetric crises, U.S. decision makers more often opt for a coercive strategy, which holds the promise of foreign policy gains without the high costs of invasion and occupation incurred by war.

For coercion to succeed, however, a target state must first yield to the demands made of it. As sovereign actors, however, states contest such external pressure; their rulers oppose concessions that would reveal them as weak and seek to avoid the humiliation of losing face for backing down. Consequently, states and their leaders resent being coerced and prefer to resist whenever feasible. Even so, the question remains as to why the United States, with its tremendous military advantage, routinely fails to coerce these weak states. U.S. leaders should understand this tension between a targeted state's desire to resist any coercive demand and its fear of America's military might. For its coercive strategies to succeed, the United States should make only those threats that it is willing to back with credible force and issue only those demands to which the target will likely concede. In so doing, it avoids foreign policy failure or the subsequent high costs of having to secure its objectives by brute force.

The U.S. record with asymmetric coercion, however, is mixed. Although the United States has adopted coercive strategies in 75 percent of its asymmetric crises since World War II, coercion has succeeded in just half of these cases (see Table 1.1). Why does the United States so frequently fail to coerce weak states? This chapter addresses this question by first considering the theoretical underpinnings of asymmetric interstate coercion. What should be the outcome of a coercive strategy adopted by a powerful challenger against a much weaker state? To accomplish this task, a theory of coercion specific to asymmetric interstate conflict is developed, key terms are defined, restrictions to the use of coercive force are examined, and alternative foreign policy options are considered. From this basis, a game theoretic model for asymmetric coercion is constructed that incorporates the strategic interaction between a powerful challenger and a weak target. From the resulting equilibrium conditions, the coercion range is formulated, the set of demands that both target and challenger prefer to resistance and brute force war. Together, the asymmetric coercion model and coercion range provide a framework to analyze why U.S. coercive strategies succeed or fail.

In equilibrium, this model expects coercion to succeed, as a great power should adopt a coercive strategy only if a coercion range exists in which a challenger and target state can reach a bargain that avoids war. In the real world, however, coercion frequently fails. To understand why weak states resist the United States, the latter sections of this chapter consider the conventional explanations for war through the lens of this asymmetric coercion model.

ASYMMETRIC CRISES

Conflicts escalate to crises between states when there is a heightened probability of military hostility accompanied by a time constraint.[1] In asymmetric interstate crises, the distribution of power is such that a great power can threaten the survival of a weaker state, but not vice versa.[2] Only a few nations since World War II have achieved great power status: the United States, the Soviet Union/Russia, Great Britain, France, and China.[3] All other states are weak by comparison. Great powers typically have the military capacity to credibly threaten to invade a weaker state and overthrow its regime.[4] Although this asymmetry is primarily determined by relative military power, other factors such as distance, geography, and climatology diffuse a great power's ability to project its military might. The disparity in power grants a great power the latitude to decide whether and, if so, when to escalate a conflict into a crisis. For example, the initial outbreak of the Bosnian Civil War did not create a crisis for the United States. Not until refugees began pouring into Western Europe and NATO peacekeeping troops deployed to Bosnia did the United States, from its point of view, make the decision to escalate the conflict into a crisis.

In an asymmetric crisis, the great power may well have the military advantage, but its interests are, by definition, limited. By contrast, the targeted state is much weaker, but its interests are greater, particularly when it perceives its survival at stake, and this translates into greater resolve.[5] The powerful state's military advantage may be in part or wholly offset by the weaker state's resolve. As a result, crisis outcomes are not determined solely by the relative level of military power but also by the willingness of the combatants to endure losses. The Vietnam War is just such an example, in which the North's willingness to endure allowed it to overcome the superior military capability of the United States.[6]

Rather than escalating a conflict into crisis, however, a great power also has the alternative of accommodating, or appeasing, a weak state, by choosing not to contest the issue at hand to avoid further conflict.[7] Given the relatively small number of international crises in comparison to the much larger universe of potential crises, accommodation is the most common foreign policy choice for all states. Accommodation is the default strategy adopted when states choose not to take military action. This is usually the reasonable course of action when nonvital interests are at stake and/or the costs of implementing a coercive or brute force strategy are too high. For example, the United States accommodated Venezuela's President Hugo Chavez after he nationalized oil rigs owned by U.S. companies in 2010.[8] Though the Obama administration sacrificed American business interests, it also avoided the economic losses and military costs of a confrontation. Great powers, in fact, most often accommodate the daily frictions generated by self-interested states forced to interact within the constraints of the international system. On occasion, however, when the issues at stake are sufficiently valuable and diplomacy alone cannot satisfactorily resolve a conflict, the United States may instead prefer to take a stand by choosing either a coercive or brute force strategy.

COERCION

In asymmetric interstate conflicts, a great power may view coercion as its most attractive foreign policy option. Coercion threatens force or employs restricted force to convince a target to comply with demands.[9] Coercive strategies consist of demands and threats, which are credibly communicated by costly signaling.

Coercive Demands: Compellence and Deterrence

Demands are the actions required of the target state to achieve the aims for which the powerful challenger initiated the crisis and, once met, terminate the crisis. Demands may be for minor policy changes, such the apology demanded of North Korea in 1975 over the murder of two U.S. Army officers in the De-militarized Zone. Demands may also be more substantive, to include regime change or relinquishing control of territory, as when the United States called for the immediate and unconditional withdrawal of Iraqi troops from Kuwait in August 1990 and insisted that Iraq not invade Saudi Arabia.

As the previous example illustrates, coercive demands can be either com-pellent, that is, for Iraq to withdraw its army from Kuwait, or deterrent, for it to refrain from invading Saudi Arabia. For our purposes, the key to differen-tiating compellence from deterrence is whether a targeted state's concessions are observable.[10] As to compellence, demands are for a target to make a change in behavior, such as "stop," "go back," "give back," or "give up." In October 1998 the United States issued two compellent demands, insisting Serbia re-duce its deployed troops to precrisis levels and allow international monitors into Kosovo. Both concessions were observable Serbian actions directly linked to U.S. demands.

By contrast, deterrent demands require the target to continue its current actions. The deterrent demand is simply "don't." With nuclear deterrence, for example, the demand is "don't launch your nuclear weapons." The causal link between a challenger's demands and target compliance is obscured in part, however, by the negative nature of the demands. The target may argue that it was not coerced, as it had never intended to launch its missiles. Deterrent demands thus provide the target state with a face-saving out that lowers its reputation costs. Reputation costs are the costs from future crises a state can expect to endure for conceding in the current crisis.[11] With deterrence, a tar-get's leader can comply with demands while claiming never to have had any plans to take aggressive action. Deterrent demands are thus relatively more palatable to the target, allowing it to more easily concede.

Coercive Threats: Punishment and Denial

Threats are the violent consequences to ensue should a weak target's response to demands not satisfy the powerful challenger. Coercive threats take two forms, punishment and denial.[12] A punishment threat is "the threat of damage, or of more damage to come."[13] If the threat is sufficiently credible, it convinces a target that it is preferable to concede now rather than to endure future pain, an approach aimed at altering the target's cost-benefit assessment. In 1999, the United States employed such a strategy against Serbia by putting at risk its economic infrastructure by the continuation of NATO air strikes, which eventually convinced President Slobodan Milosevic to concede control of Kosovo.

In lieu of punishment, a great power may choose to coerce by denial.[14] Here, a challenger attacks a target's ability to defend the objective at stake. If the great power can convince a weak state that the situation is hopeless, the weak state then has the incentive to concede to avoid further losses from futile fighting. For example, on February 23, 1991, after six weeks of intense air strikes against Iraqi forces, Saddam Hussein conceded to a Soviet proposed peace plan to withdraw his troops from Kuwait. Although the United States refused to accept Iraq's concessions, the coalition's air strikes had denied Iraq both the means and the hope of defending Kuwait against the impending invasion.[15]

Whereas a punishment strategy threatens that which the target values, a denial strategy lowers the target's probability of victory in war by altering the balance of military power. Denial strikes are directed against an enemy's defenses and, as such, may require a considerable expenditure of force. Should the conflict involve an issue the target highly prizes, a denial strategy can prove nearly as costly to the challenger as an invasion and, in its initial and intermediate stages, may well be indistinguishable from a brute force strategy.[16] However, whereas brute force succeeds when an objective is seized, denial succeeds when the enemy concedes the objective without the challenger having to incur all the costs and risks of invasion and occupation.

Costly Signals

A powerful challenger may make threats it does not intend to keep in an effort to gain larger concessions. A weak state, aware of this incentive, is likely to discount such threats as cheap talk.[17] In 1946, when the Soviet Union demanded the right to naval bases in the Turkish straits, Turkey, backed by the U.S. and British navies, called Stalin's bluff, and the Soviets eventually backed down. To overcome skepticism that its threats are credible, a great power must employ costly signals to demonstrate its willingness to incur costs that a less resolved challenger would not endure.[18] During the Christmas bombings of Hanoi in 1972, the United States did not relent in its attacks even though it incurred the loss of twenty-six aircraft, including fifteen of its nuclear-capable B-52 bombers, in just eleven nights of bombings. This demonstration of U.S. resolve succeeded in bringing the North Vietnamese back to the negotiating table to sign the Paris Peace Agreement.[19]

Costly signals are sunk costs, which are incurred regardless of a target's response to them. Costly signals are designed to separate the resolute from the bluffer, reducing a target's uncertainty as to the credibility of a great power's threats. Signals range from exemplary military actions to the restricted use of force. Exemplary actions, such as naval exercises or troop deployments, demonstrate a great power's capability and interest while keeping operational expenses relatively low. For instance, on October 8, 1994, in response to the movement of Iraqi Republican Guard units toward the Kuwaiti border, the Pentagon announced that it had alerted two air combat wings, issued deployment orders for 4,000 soldiers, and ordered ships containing the heavy weapons and equipment for another 15,000 troops to the Persian Gulf.[20] In response to this overt signal of U.S. resolve, Iraq quickly withdrew its forward units without a shot being fired.

Signaling costs increase exponentially, however, once the threshold of violence is crossed. Military actions, such as the limited NATO air strikes against Serbia in 1999, carry not only greater operational costs but also the risk of combat losses and conflict escalation.[21]

Coercion Success or Failure

A state engages in coercion to obtain its foreign policy objectives. The degree to which these aims are achieved determines whether coercion is a success or a failure. This assessment can be made by a comparison of the challenger's ex ante objectives with the final outcome. Coercion is deemed a success if the challenger achieves its core objectives and a failure if it does not. Core demands are typically those closely linked to the issue that initiated the conflict. For instance, the United States demanded of Libya in 1991 that it extradite two suspects in the bombing of Pan Am Flight 103. The United States later relaxed an additional requirement that the two be remanded to U.S. or British custody, eventually agreeing instead to their extradition to The Hague.

Requiring the challenger to achieve all of its demands, however, is too strict a standard for coercion success. As mentioned earlier, given the uncertainty over interests, capabilities, and resolve, a challenger has an incentive to make greater demands to gain a better bargained outcome. Further, diplomats realize that, in the course of negotiations, they will likely need to concede on some points to gain others. By compromising, a challenger can provide a target's leader with the means to save face despite making concessions. For example, in response to North Korean soldiers killing two American officers at Panmunjom in August 1976, the U.S. State Department demanded North Korea accept responsibility for the killings, make assurances none would happen again, and punish those responsible.[22] The crisis concluded when U.S. leaders accepted a letter from Kim Il-sung that expressed regret for the incident yet did not take responsibility for the killings nor indicate that those responsible would be punished. However, given the regretful tone and singular nature of Kim Il-sung's letter, it being the first such correspondence ever received directly from the dictator, the United States chose to overlook its remaining demands in favor of a resolution to the crisis which de-escalated tensions on the peninsula.

In addition to a target not conceding to demands, coercion also fails if the challenger abandons it for a brute force strategy. Coercion can succeed only if both the target and the challenger accept the coercive outcome. For example, in February 1991 the United States did compel Saddam Hussein to accept a Soviet peace plan. Coercion failed, however, as the United States rejected the conditional terms of the proposed cease-fire. Thus a crucial measure of coercion success is whether concessions by the target result in crisis termination.[23]

Finally, one should note that coercion success or failure is based on effectiveness rather than efficiency. Effectiveness measures whether objectives are achieved, whereas efficiency is a relative comparison of the costs and benefits of the strategy adopted against viable alternative strategies.[24] Postcrisis analysis of the efficiency of coercive strategies is important for the United States to learn the limitations of coercion and use that information to update its expectations for future conflicts. The United States Strategic Bombing Survey (USSBS) of the European Theater in World War II is an excellent example of just such analysis. The survey evaluated the Allied strategic bombing campaign and noted that, while the Combined Bomber Offensive significantly decreased German production, the bombing of civilians to lower morale did not significantly lower worker productivity. The USSBS concluded that strategic bombing would have been more efficient had it instead used more of its resources against Germany's energy sources, in particular against its electrical grid.[25] Unfortunately, such fine-grained data are not available for most crises. The USSBS reports were generated only after the occupation of Germany and by the work of over a thousand civilian and military analysts. Because such data to evaluate efficiency are rarely available, this book will focus instead on assessing coercion effectiveness.

Restrictions to Coercive Force

Coercion may entail not only threats of force but also the actual employment of restricted force, whereby the purpose of violence is to signal the credibility of a challenger's capability and resolve. Yet to be defined are the parameters for this "restricted" force, that is, where is the threshold between coercive and brute military force? The answer, in part, depends on whether the coercive threat is one of punishment or denial. For punishment strategies, force is restricted to the minimum required to credibly communicate the loss that is yet to come should demands not be met.[26] Additional strikes are actually counterproductive, as they reduce the number of targets available to threaten in later strikes. This logic is best remembered for its use by the Johnson administration during the early days of the Rolling Thunder air campaign in 1965, in which restraints were placed on the most lucrative North Vietnamese targets to hold them hostage to future attack. More successfully, the 1999 NATO air campaign against Serbia initially restricted attacks on its economic infrastructure, only to later escalate air strikes on such targets to ratchet up the pressure on Serbian leaders.[27]

In contrast to punishment, a denial strategy employs sufficient force to demonstrate both the great power's military capability and its resolve until the target gives up hope of winning. Denial attacks are aimed against a target's military and not its population. The coalition air campaign of 1991 predominantly targeted Iraqi military forces and pressured Saddam to accept the Soviet peace proposal.[28] Such a restricted air campaign represents the upper range of military force, that is, operations short of ground invasion, which are considered to be coercive.[29] Although the threat of an invasion is coercive, the commencement of a major ground offensive is more appropriately viewed as a brute force strategy.

Coercive Diplomacy

Coercive diplomacy is a subset of coercion that places further restrictions on military action, reducing its employment to "exemplary or symbolic use."[30] Coercive diplomacy relies on the threat but not the actual use of force. It fails once violence is employed. Actions such as elevating alert levels, mobilizing or deploying forces, or military exercises signal the credibility of a challenger's threat without actually engaging in violence. The intent of coercive diplomacy is to "back a demand on an adversary with a threat of punishment for non-compliance that will be credible and potent enough to persuade him that it is in his interest to comply with the demand."[31]

A strategy of coercive diplomacy is not limited to threats only but may also be coupled with positive inducements as part of a stick-and-carrot approach.[32] The United States succeeded at coercive diplomacy in 1994 when North Korea agreed to freeze its nuclear program in the face of military threats combined with the promise of light water reactors and fuel oil.[33]

Economic Sanctions and Coercion

Economic sanctions are a foreign policy option that can be used either as a substitute for or a complement to a coercive military strategy. Since the days of Thucydides and Pericles's Megarian decree, great powers have employed trade, financial, and weapons embargos as punishment strategies to increase the costs for resisting or as denial strategies to weaken a target's defenses. Economic sanctions have been criticized as causing needless civilian suffering while achieving little, particularly when matched to demands for regime change.[34] Less conclusive, however, is the effect of sanctions in a complementary strategy, in which their use is only indirectly reflected in the outcome of military operations. For instance, the value of the arms embargo against Iraq during the 1990s could not be evaluated until Iraq's reduced military capability was revealed in 2003.

The United States often employs sanctions against weak states, and sanctions have proven a key coercive tool in several crises examined in detail in this book, including the Bosnian Civil War and Libya's abandonment of its WMD. Sanctions, however, are excluded as a separate strategy in the asymmetric coercion model for three reasons. First, as already discussed, sanctions are often employed in mixed strategies and can thus be treated as part of the overall coercive strategy. Second, sanctions are used at times to mask the real underlying strategy of accommodation, as U.S. leadership imposes ineffective targeted sanctions designed primarily to silence domestic and international criticism. In July 1999, President Clinton issued an executive order for targeted sanctions against the Taliban in Afghanistan, freezing their assets in the United States and boycotting the import and export of most goods and services.[35] Because the Taliban had few financial holdings and conducted little trade with the United States, these sanctions had an insignificant impact on Afghanistan policy. They did, however, allow the Clinton administration to deflect criticism. Finally, economic sanctions are excluded as a separate strategy within the asymmetric coercion model because the underlying interaction between the powerful challenger and weak target is driven by the threat of violence and not the threat of economic suffering. That being said, the logic of economic coercion closely parallels that of military coercion. The conflict with Libya over the Pan Am Flight 103 bombing is included in Chapter 6 to assess how the asymmetric coercion model can also be used to analyze cases that do not entail threats of force. Also, because sanctions play an important role in the U.S. crises with Iraq, Serbia and Libya, their impact will be discussed, where applicable.

MODEL OF ASYMMETRIC COERCION

Having introduced and defined terms for asymmetric coercion, we now turn to the interaction between a great power and a weak state and develop a model to explain their strategic decision making. Central to this model is the premise a great power does not have its survival threatened in an asymmetric conflict and therefore has the latitude to escalate a conflict to crisis and, when it chooses to coerce, to vary the demands and the threats that it makes. This powerful challenger optimizes its expected outcome by balancing its demands and threats, considering the impact of threat level when choosing demands and vice versa.[36]

For an example of this shifting of demands and threats, consider U.S. actions against Serbia in 1999. In February, the United States demanded the Serbs remove most of its forces from Kosovo, allow in NATO troops as peacekeepers, and, in three years' time, hold a conference on the future independence of Kosovo.[37] In June, the Clinton administration lowered these demands by offering to have the peacekeepers report to the UN Security Council, by allowing in Russian troops, and by omitting any reference to Kosovo independence. Simultaneously, NATO ratcheted up its threats. In February it threatened a restricted three-day air campaign aimed primarily against Serbian military targets. By June, NATO air strikes had escalated to directly target Serbia's economic infrastructure. Over time, the United States lowered its demands while increasing its threats until it convinced Serbia to agree to its terms.

Asymmetric Coercion Assumptions

Theories of international politics require simplifying assumptions as to the nature of the international environment and the actors therein. This theory begins with neorealist assumptions of an anarchic, self-help, international system with states as the primary actors.[38] Given certain constraints and options, state leaders make decisions they believe will result in an "optimal" outcome. The adoption of a rational actor framework, however, does not imply that psychological, cognitive, or group/organizational biases are unimportant to decision making.[39] In fact, these factors are relied on to help explain why some U.S. coercive outcomes differ from the model's expectations. Although states are initially assumed to be unitary actors, in the next chapter the assumption is relaxed on the targeted state to also consider the interests of its ruling regime. This incorporates domestic power considerations that frequently alter a target state's decision calculus.[40]

The Asymmetric Coercion Model

A conflict arises between a powerful challenger and a weak target over an issue controlled by the target (see Appendix B for the formal model).[41] Issues range from relatively minor matters, such as a target's policy toward an ethnic group, to larger issues over control of its territory or government. Again, the case of Kosovo illustrates this range of issues. In October 1998, the conflict issue focused on Serbia's domestic policies toward its ethnic Albanian population. By February 1999, the issue had escalated to territorial control of Kosovo.

States gain by controlling as much of the conflict issue as possible. The marginal value for both states is positive but with diminishing returns to scale. For example, in the 1991 Gulf War, maintaining control of southern Iraqi territory proved to be less important to the United States than liberating Kuwait. Conversely, although Saddam Hussein agreed to the Soviet peace plan to withdraw Iraqi forces from Kuwait in February of 1991, he never conceded to the forfeiture of Iraqi sovereignty over the northern safe area or the southern no-fly zone. Saddam valued Kuwait, but he treasured Iraq more.

A challenger adopts the strategy with the best expected outcome from the foreign policy options available to it: accommodation, brute force, or coercion (see Figure 2.1). It can opt to accommodate, and the conflict ends without a crisis. If it chooses to do so, the great power gains nothing while it incurs a reputation for appeasing the weak state.[42] These reputation costs are losses a state can expect to suffer in future conflicts for demonstrating a lack of

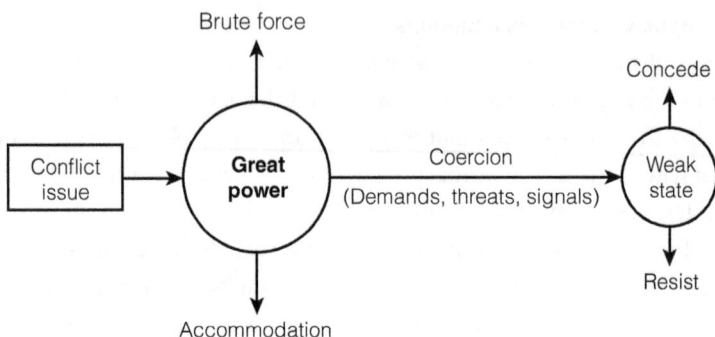

Figure 2.1. Asymmetric coercion model.

resolve today.[43] If either coercion or brute force has a better expected outcome, the challenger does not accommodate but instead escalates the conflict into a crisis.

BRUTE FORCE STRATEGY

As an alternative to accommodation, the challenger may instead seize its objective by force. The expected outcome for a brute force war depends on the value it places on the issue at stake, discounted by the probability of victory. Though the United States has overwhelming force, its probability of victory need not equate to the balance of power between it and a weak target state. Military power can be diffused by distance, terrain, and climatology and is in part offset by the higher resolve of a target state. For example, in September 1950 the United States responded to the North's invasion of South Korea with a brute force strategy culminating in the amphibious landing at Inchon, a move that cut the lines of communications and thus decisively defeated the North Korean Army. This successful brute force outcome was far from certain, however, as the United States had to first project its military power halfway around the world and then depend on a risky and complex amphibious operation for victory.[44]

In addition to calculating the probability of a brute force victory, the challenger must also factor in its expected costs of invasion and occupation. As the United States has experienced in Afghanistan and Iraq, the costs of occupation can be significantly higher than those of the invasion. Despite the fact that a brute force strategy can obtain larger aims such as homeland territory or regime change, its high price tag usually means that the expected outcome for brute force war is less than that of either accommodation or coercion.[45]

COERCIVE STRATEGY

Rather than brute force, then, a powerful challenger may choose coercion. When it does so, it issues a coercive offer consisting of a demand backed by a threat, the credibility of which is communicated by a costly signal. In response, the weak state either concedes or resists.[46] If it concedes, the expected outcome for the target is the amount of the conflict issue it can retain less any reputation costs incurred for backing down.[47] An example of such reputation costs can be found in Qaddafi's 1998 agreement to hand over two officials to stand trial for the bombing of Pan Am Flight 103. Libya's concession led only to the United States making further demands that Libya also abandon its chemical and nuclear weapons programs.[48]

Should the weak state initially resist, however, the outcome of the crisis depends on the likelihood a powerful challenger will enact its threats and whether these actions will then convince the target to cede to demands. The likelihood that coercion will then succeed can be thought of as a lottery, with the odds determined by the probability that the target will ultimately concede once the challenger makes good on its threats. In the crisis over Kosovo, Serbia resisted signing the Rambouillet Accord, which led NATO to enact its threatened air strikes. Though Milosevic was initially obstinate, air strikes eventually convinced the Serbians to concede control of Kosovo.

In addition to the benefits gained by resisting, the target state's expected outcome must also incorporate its costs for resisting. These are the losses the target suffers when the challenger carries out its threats. For punishment strategies, these are the economic and civilian losses from punitive strikes. For Serbia the most critical costs were the economic losses it suffered as the war continued. Under denial strategies, these costs are measured by the degradation of a target state's military capacity and the subsequent weakening of its defenses.

Turning to the powerful challenger's calculations, its expected outcome for coercion success is the value it places on its objective minus any signaling costs. In 1994 the United States convinced Iraq to withdraw elements of two Iraqi Republican Guard divisions deployed along the Kuwaiti border. U.S. leaders thus compelled Iraq while incurring only the minimal signaling costs of deploying additional forces to the region.

Should the target resist despite signaling, the expected benefit to the challenger depends on the likelihood that coercion will succeed once it makes good on its threats. On the other side of the ledger, the costs include those signaling costs already incurred plus the expected costs of enacting its threats. With Kosovo, the United States incurred approximately $20 billion in operational expenses to do so.[49]

When a weak state resists and a great power makes good on its threats, then the crisis may escalate into war, as with Kosovo in 1999 and Iraq in 1991 and 2003. However, in this book, to retain the distinction between coercive and brute force strategies, only brute force will be referred to as *war*.

OPTIMIZING COERCIVE DEMANDS AND THREATS

Having evaluated the expected outcomes for target concession or resistance, now consider how a rational challenger would choose the optimal coercive offer to maximize the demand made at the lowest credible threat level. Such an offer minimizes the risk that the target will resist and therefore avoids the costs of having to carry out threats (see Appendix B). Assuming that a great power is a utility maximizer does not suggest that its leaders always behave rationally. Rather, the model provides a basis to begin assessing coercive outcomes in real-world cases.

Returning to the model, the optimal demand consists of the highest demand for which the target is indifferent as to conceding or resisting.[50] This is the equilibrium condition for the model because an increase in demand would cause the target only to resist and coercion to fail. Calculating this demand, however, is only half of the challenger's optimization problem. The demand must also be backed by a threat, and signaling the credibility of that threat is costly. The challenger's problem therefore entails issuing the optimal demand at minimum signaling costs. Given the decreasing marginal benefits for the conflict issue, along with the increasing marginal costs for signaling, an interior solution exists whereby the challenger's optimal offer limits both the demands and the threats made. The intuition is that a powerful challenger gains by increasing its demands, but, as previously discussed, the marginal benefits from additional demands accrue at a diminishing rate. The threats to back these larger demands must likewise be raised, which to be seen as credible requires additional signaling costs. Unlike benefits, the marginal cost of signaling accrues at an increasing rate, particularly when military action crosses the threshold of violence. Exemplary uses of force such as military exercises, deployments, or mobilizations are costly but not nearly as much as combat operations, with the risks of battlefield casualties and conflict escalation. As a result, the challenger optimizes by increasing its demands until the marginal benefits are offset by the increase in the marginal costs of signaling. If no interior solution exists, the challenger should not coerce but instead select the upper bounded solution of brute force war.[51]

THE COERCION RANGE

When a challenger chooses coercion, there is a set of demands that includes the optimal demand previously discussed, which the great power prefers over the alternative of a brute force war. Conversely, though a target should concede to the challenger's optimal demand, it prefers any concession less than that. The coercion range lies where these demands and concessions overlap.[52] A normalized line segment in Figure 2.2 represents the conflict issue, with the left end point, labeled "0," being the outcome in which the challenger accommodates and receives nothing and the target retains the entire issue. The right end point, labeled "1," represents the opposite outcome in which the challenger obtains the entire conflict issue by brute force. For conceptualization, consider the conflict issue to be territory the target state controls. The coercion

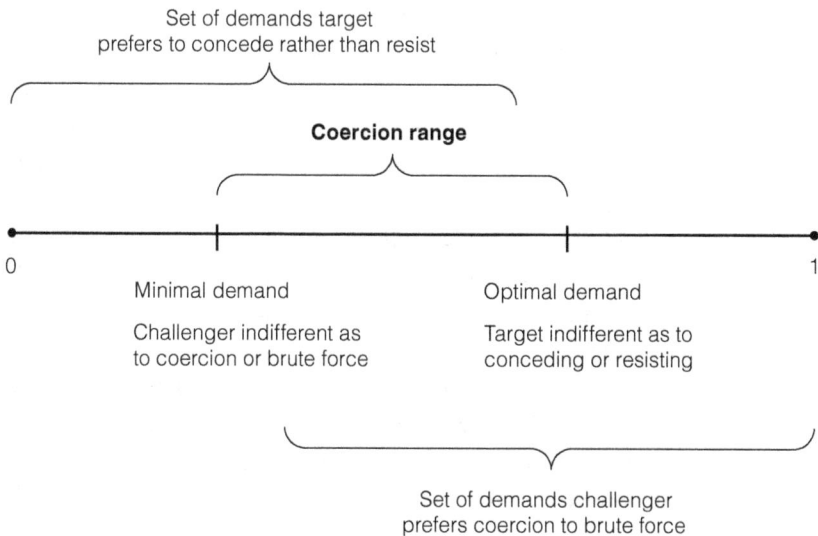

Set of demands target
prefers to concede rather than resist

Coercion range

0
Minimal demand

Challenger indifferent as
to coercion or brute force

Optimal demand

Target indifferent as to
conceding or resisting

1

Set of demands challenger
prefers coercion to brute force

Figure 2.2. Coercion range.

range's lower bound is the minimal concession the challenger will accept in lieu of brute force war, and the upper bound is the optimal demand, the maximal demand to which the target will concede (see Appendix B).

If the minimal concession that the challenger is willing to accept exceeds the maximal demand for which the target will concede, then no coercion range exists, and the challenger should instead adopt a brute force strategy.

Explaining Coercion Failure

The game theoretic model for asymmetric coercion derives the optimal demands, threats, and costly signals for a powerful challenger's coercive strategy. The coercion range illustrates the demands that the powerful challenger prefers to war and the concessions that the weaker target state prefers to resistance. In equilibrium, coercion should succeed, as a challenger should choose a coercive strategy only when the target is likely to concede.

Yet, in the real world, weak states frequently resist great powers. Crises often conclude as foreign policy failures, as in 1986 following the El Dorado air raid against Libya, which did not change Qaddafi's policies regarding terrorism. Or crises may be solved only once coercion fails and the United States switches to a brute force campaign, as with the invasion of Iraq in 2003. The next section examines the conventional explanations for war, that is, why coercion fails.

Explanations for Coercion Failure and War

Thomas Schelling commences *Arms and Influence* by declaring diplomacy to be bargaining and the threat of violence to be bargaining power.[53] Such strong-arm bargaining, that is, coercive diplomacy, more often than not fails, however, with states resorting to violence to resolve conflicts.[54] Many international relations scholars have viewed such wars as the result of mistakes made by states, regimes, and leaders who do not always act rationally. Due to psychological and cognitive biases, organizational and bureaucratic politics, illness and age, time constraints, different time horizons, and varying degrees of risk acceptance, decision makers misperceive and/or miscalculate their opponent's capability and resolve (as well as their own). In terms of the asymmetric coercion model, nonrational factors cause the challenger and target to draw different conclusions regarding the expected outcome of coercion and brute force war. This can lead a great power to mistakenly choose coercion over the alternatives of accommodation and brute force. In addition, the great power may then mismatch its demands, threats, and costly signals.

No coercion
range exists

0   1

Optimal demand   Minimal demand

Target's maximum   Challenger's
concession   minimum demand

Figure 2.3. Coercion failure due to nonrational behavior.

This can ultimately lead to coercion failure and war if the weak state's maximum concession is less than the great power's minimal acceptable demand (see Figure 2.3).

An example of coercion (and coercive diplomacy) failure due to a target state mistakenly assessing its maximum concession to be less than the challenger's minimal demand is Saddam Hussein's rejection of UN resolutions demanding the unconditional withdrawal of Iraqi troops from Kuwait. Prior to the commencement of air strikes on January 17, 1991, Saddam significantly underestimated the balance of power between Iraq and the U.S.-led coalition. He also misperceived American resolve, as he expected that the United States would not tolerate the casualties he anticipated from a war. Saddam miscalculated the probability of victory and costs for fighting a brute force war. As a result, the maximum concession he would entertain prior to the start of the fighting linked the resolution of the Kuwaiti crisis to the Israeli–Palestinian conflict. This concession was less than the minimal demand the United States would consider.

Not only are weak states susceptible to misperception, but great powers make mistakes too, not only in their choice of a coercive strategy but also in its execution. In the 1980s the White House demanded that Libya curtail its policy of supporting international terrorist organizations. When evidence implicated Qaddafi's regime in the bombing of a Berlin discotheque, which killed U.S. service members, President Reagan ordered air strikes on Qaddafi's compound in April 1986. Coercion failed to convince Libya to change its policy, however, as the Reagan administration could not maintain a credible threat of force. International and domestic pressure convinced Reagan not to authorize strikes without further "smoking gun" proof of direct Libyan involvement. Qaddafi, however, restrained his rhetoric and Libya's overt support of terrorist attacks, thus reducing the credibility of U.S. threats of force. Unfortunately, Libya covertly continued its terrorist policies, as evidenced by the bombing of Pan Am Flight 103 in December 1988.

In addition to nonrational explanations for coercion failure, other scholars have pointed to uncertainty and private information being responsible for war.[55] If challenger and target are privy to different information, even if both are rational, they may reach differing assessments as to the other's interests, capabilities, and resolve. If these estimates vary, such that the challenger's range of acceptable demands does not overlap with what the target is willing to concede, then coercion fails. Graphically, the absence of a coercion range is similar to that previously illustrated in Figure 2.3 for nonrational coercion failure.

In theory, states could avoid war by revealing to each other the information they keep private. There are, however, incentives for the challenger and the target to bluff and misrepresent their intentions and capabilities to obtain a better expected coercive (bargained) outcome, even if this means accepting the risk of war.[56] There is also a reason to keep capabilities and intentions secret because revealing them to an opponent may decrease the probability of victory in a subsequent war. In the seminal article "Rationalist Explanations for War," James Fearon argues this to be the problem Japan faced with Russia in 1904. The Russians significantly underestimated Japanese military and naval power. The Japanese, however, could not reveal their power without compromising their war plans, which heavily depended on the element of surprise.[57]

States may therefore be left with violence as the only means available to obtain information as to the capability and resolve of their opponent, the true balance of power, the probability of victory, and the costs of fighting. As Geoffrey Blainey writes in *The Causes of War*, "War itself then provides the stinging ice of reality. And at the end of a war those rival expectations, initially so far apart, are so close to one another that terms of peace can be agreed upon."[58] A great power, however, may not have to wage a brute force war to update its assessment of the conflict. A coercive strategy's restricted use of violence may better serve the purpose of revealing information to both the challenger and target to reach a negotiated outcome short of war. Returning to the 1991 Iraq example, after the Iraqi Army had been under air attack for five weeks, Saddam had a new appreciation for U.S. military power. He would now accept a cease-fire in exchange for the withdrawal of his forces from Kuwait.

There are two limitations with a rational explanation for war due to uncertainty and private information and nonrational explanations based on misperception and miscalculation. First, these factors are present in all crises. These explanations are useful for understanding why states do not reach a peaceful negotiated settlement for a specific crisis, but they do not provide ex ante predictors for whether a coercive strategy is likely to succeed or fail. Second, as Blainey points out uncertainty and nonrational behavior better explains the onset of war, that is, why coercive diplomacy fails, but does not explain why a war may continue even after states have updated their information and reevaluated the expected outcome of coercion and war. Now consider two additional arguments why states may rationally choose war even when they both have complete information.

One structural explanation for war blames the anarchic nature of the international environment.[59] Given the lack of a hierarchical power to enforce agreements, a commitment problem occurs where a powerful challenger cannot credibly promise to make good on its obligations in the future when the agreement has been implemented.[60] This is problematic for a great power that would ex ante prefer a negotiated settlement that avoids the costs of war but cannot credibly promise to refrain from making more demands once a target state concedes. Weak states may believe that concessions today will lead only to further coercive demands by the challenger tomorrow. As a result, the target takes a stand now rather than later, when it will be in an even more disadvantageous position.

There are two explanations for why the weak state may have such a pessimistic view of future conflict. First, material concessions now may tilt the balance of power further to the great power's advantage, increasing its probability of victory and thus improving the challenger's expected outcome from a brute force war. This then introduces an incentive for the great power to introduce new demands. Second, instead of a shift in the balance of power, a target's concessions may instead reveal the target state as unresolved. This also introduces an incentive for the challenger to initiate another crisis. This has the impact of increasing the target's reputation costs for making concessions in the present crisis.[61] If these costs are sufficiently high, the coercion range may be eliminated.

In practice, however there are ways the United States ameliorates commitment problems, as will be demonstrated in Chapters 4 to 6. The formation of broad international coalitions, such as the U.S.-led coalition during the Gulf

War, increased the potential diplomatic costs had the United States later decided to invade Iraq. The inclusion of other great powers in negotiations, such as the Soviet Union/Russia's involvement with Iraq in 1991 and Serbia in 1999 likewise increased U.S. diplomatic costs had it reneged on its promises; as a result, this lessens commitment worries. Finally, the implementation of tit-for-tat concessions can incrementally produce trust even between enemies, as the United States and Libya demonstrated by eventually resolving their conflict over the Pan Am Flight 103 bombing.

A third rational explanation for war is that of issue indivisibility. It asserts that there are certain issues over which a state is unwilling to negotiate, preferring resistance to any peaceful settlement.[62] Fearon acknowledges issue indivisibility as a theoretically viable rationalist explanation for war but dismisses it as inconsequential for modern international politics. He argues, though does not provide evidence, that "issues over which states bargain typically are complex and multidimensional; side-payments or linkages with other issues typically are possible . . . War-prone international issues may often be *effectively* indivisible, but the cause of this indivisibility lies in domestic political and other mechanisms rather than in the nature of the issues themselves."[63] Robert Powell takes this argument even further by claiming that issue indivisibility is, in fact, no more than a commitment problem.[64]

Those who do claim issue indivisibility as a rational cause of war point to specific religious sites or attributes of a particular territory as integral to a nation's identity, and therefore the states do not view these issues as being divisible.[65] Ron Hassner, in *War on Sacred Ground*, asserts certain religious sites possess unique, bounded, and cohesive properties, making them systemically indivisible.[66] These properties are roughly analogous to the characteristics of a discrete economic good. Viewed in terms of the coercion range, the target state views the issue as indivisible, and there is therefore no concession it is willing to make. For instance, Israel views the Temple Mount in Jerusalem as nonnegotiable.

Regardless of whether issue indivisibility is more logically considered a separate explanation for coercion failure or viewed as a commitment problem, the number of U.S. asymmetric crises, in which issue indivisibility is evident, is relatively small. Only in the crisis over Kosovo, the historic birthplace of Serbia, does the issue of indivisibility appear to be relevant (see Chapter 5). And even then Milosevic eventually conceded the territory.

All these conventional rational and nonrational explanations for war are limited in that they do not fully explain why coercion either succeeds or fails. The next chapter introduces an alternative explanation for why weak states resist great powers based on survival concerns.

CONCLUSION

Crises between the United States and weak states are recurrent, and U.S. leaders more often resort to coercion over brute force war. Unfortunately, coercion works only half the time, leaving the United States to either accept a foreign policy failure or resort instead to brute force war. This chapter developed an asymmetric coercion model to explain the dynamics of a great power coercing a weak state. The model's equilibrium conditions demonstrated that, theoretically, coercion should always succeed; a powerful challenger should adopt a coercive strategy only when the target state is likely to concede. To further illustrate this idea, the coercion range illustrated the room for compromise available to a powerful challenger and a weak target. The second half of this chapter returned to consider why a weak state might still resist a great power by reviewing the nonrational explanations for war based on misperception and miscalculation along with rational explanations due to uncertainty and private information, commitment problems, and issue indivisibility. The next chapter introduces survival concerns as a reason weak states will resist the United States. In addition, the coercion model's unitary actor assumption is relaxed and reasons are considered why a weak state's leaders might not concede to coercive demands. Finally, the logical follow on question is considered as to why U.S. decision makers might adopt a coercive strategy even when it is likely to fail.

3 SURVIVAL AND COERCION FAILURE

ON MARCH 17, 2003, PRESIDENT GEORGE W. BUSH demanded that Saddam Hussein and his sons depart Iraq within forty-eight hours and that the Iraqi military permit U.S. forces to enter Iraq to rid it of its weapons of mass destruction.[1] U.S. coercion failed, however, as Saddam ignored the ultimatum, refusing to concede to an imposed regime change and occupation by foreign troops. Anticipating such a response, National Security Advisor Condoleezza Rice had instead favored an alternative draft of Bush's speech in which the pending invasion would simply be announced.[2] Rice understood that the threat of invasion was insufficient to convince Saddam Hussein to agree to such draconian terms.

This chapter examines why the United States would intentionally make high-level coercive demands likely to be rejected by weak target states. Demands for homeland territory or regime change threaten the survival of a weak state and its regime and are likely to be resisted so long as it has the means to do so.

In the previous chapter, asymmetric coercion was defined as a powerful challenger's threat of force or the restricted act of force to convince a weak target to comply with its demands. When considering a coercive strategy, a great power also has the option of either accommodating the weak target state or seizing its objective by brute force. The asymmetric coercion model incorporates this strategic choice and the interaction between the powerful challenger issuing demands backed by threats, communicated through costly signals, and the weak target deciding whether to concede or resist. Theoretically, asymmetric coercion should always succeed, as a great power should make only those demands to which a weak state will concede. From this equilibrium condition the coercion range was developed, the set of demands for which the challenger prefers coercion to brute force and the concessions the target prefers to resistance. Together, the asymmetric coercion model and coercion range provide a framework for analyzing asymmetric crises and for policy makers to employ when developing coercive strategies.

Although in theory asymmetric coercion should always succeed, in practice it often fails. Since World War II U.S. coercive strategies have succeeded only half the time. The last chapter considered the existing literature on conventional explanations for coercion failure. For a variety of reasons, a state's leaders can misperceive and miscalculate and fail to negotiate a peaceful settlement. Even if states act rationally, they may not be able to overcome uncertainty or commitment problems that, given the anarchic nature of the international system, may also lead to war. None of these explanations for war, however, adequately clarifies why a great power would knowingly choose a coercive strategy expected to fail. Why did the Bush administration make such a coercive demand in 2003 that directly threatened the survival both of Saddam's regime and of Iraq itself?

This chapter introduces a three-part explanation for why great powers often fail to coerce weak states. Paradoxically, asymmetric coercion may fail because of the very disparity in power that defines it, an imbalance that induces a powerful challenger to make high demands of a weak target. Should a great power choose to escalate a conflict into a crisis, its options are those of coercion or brute force war. Because a great power's probability of a brute force victory against a weak state is likely to be high and its costs of fighting low, its expected outcome from such a war is relatively large. For a great power to instead prefer coercion, the concessions made by the target state must also be large. The greater the imbalance of power and the expected outcome of a brute force strategy, the higher the corresponding coercive demands must be to compensate the challenger for not simply taking its objectives by force. Unfortunately, the higher these demands are, the greater the chance that they threaten the target's survival. For example, following Iraq's invasion of Kuwait in August 1990, the United States demanded that Iraq immediately withdraw its troops from Kuwait. As the U.S.-led coalition began to deploy forces to the region, the probability of its victory in a war with Iraq increased. As a result, the Bush administration expanded its objectives beyond the liberation of Kuwait to also include the destruction of the Iraqi Republican Guard, a concession Saddam would not make as he depended on the army to defend his regime and the homeland.

The second part to this argument identifies the sovereignty concerns that threaten a weak state's survival. A critical domestic sovereignty issue is a state's control of its homeland territory.[3] Only on rare occasions have states agreed to relinquish portions of their homeland, such as Czechoslovakia surrendering the Sudetenland and Serbia relinquishing Kosovo. Even in these cases, however, these weak states conceded only portions of their homeland in an effort to preserve the remainder of their state. Just how much territory a state can lose and remain viable is beyond the scope of this book. States do require land, however, and a nation's borders make a natural and observable demarcation beyond which a targeted state is not likely to negotiate.

In addition to homeland territory, regime change also threatens state survival. International sovereignty over foreign policy decision making is a defining characteristic for all states in the international system.[4] The forfeiture of such autonomy through an externally imposed regime change results in a state's loss of control over its foreign policy, a condition Tanisha Fazel has termed *state death*.[5] Though its chances of victory against a great power may be small, a weak state will prefer to resist and fight rather than concede to certain demise. Fortunately, since the end of World War II state death has been rare. When it does occur, however, it can be extremely costly not only for those defeated but also for allies and enemies alike, as the United States discovered in Vietnam and Iraq.[6] Though state death is infrequent, the *threat* of state death is prevalent, particularly in asymmetric international crises where great powers have the military capability to invade and occupy weak states. The recurrence of asymmetric crises and the extreme stakes at play for weak states make the threat to state survival a relevant, structural explanation for coercion failure.

In summary, asymmetries in power induce great powers to levy heavy demands for a weak state to forfeit its sovereign control over its territory or its regime. Such demands, however, threaten the survival of the weak state causing it to resist and coercion to fail. The following sections examine in more detail this argument for why great powers make such high demands and why weak states reject them. The final piece to the puzzle, however, is to explain why a great power would then knowingly adopt a coercive strategy likely to fail. Why would the United States incur the costs of signaling a high-level coercive demand rejected by the weaker state? Why would it not instead move directly to a brute force strategy? The end of this chapter introduces two explanations for why U.S. leaders may choose coercion when they expect, and at times even desire, such a strategy to fail.

WHY GREAT POWERS MAKE HIGH-LEVEL COERCIVE DEMANDS OF WEAK STATES

The asymmetric coercion model developed in the previous chapter provides a framework for analyzing why a great power may make a high-level demand of a weak state. A powerful challenger escalates a conflict into a crisis by threatening or using force in either a coercive or brute force strategy. The great power optimizes the coercive outcome by making the largest demand, backed by the lowest credible threat made with the least costly signal to which a weak target is just willing to concede. This procures for the great power the greatest gains while avoiding the costs of having to carry out its threats. The expected outcome from a brute force strategy, by contrast, depends on a great power's probability of victory and costs of fighting. The coercion range depicts the set of coercive demands that both the challenger and the target prefer to a brute force war (see Figure 2.2).

In a brute force war, the probability of victory is determined by several factors, a key one being the balance of military power between the warring parties. In asymmetric interstate conflicts this balance is, by definition, disproportionately in favor of the great power and, ceteris paribus, its probability of victory is likely to be high.[7] The high expected outcome for a great power in a brute force war may, in turn, eliminate the coercion range (see Figure 3.1). The paradox lies in the fact that the large asymmetry in power makes it less likely that any successful coercive outcome exists.

In the crisis between the United States and Afghanistan following Al Qaeda's September 11, 2001, attacks, no coercion range existed due, in large part, to the extreme imbalance of power. The United States vastly surpassed Afghanistan in every conceivable measure of power: demographic, diplomatic, economic, and military. The probability of U.S. victory against Afghanistan thus was high and cost of fighting relatively low. When the Taliban ignored U.S. demands to hand over Al Qaeda's leaders, the Bush administration

Coercion range
eliminated

⟶

0   1

Target's   Challenger's
maximum   minimum demand
concession   increases with an
                              increase in the
                              probability of victory

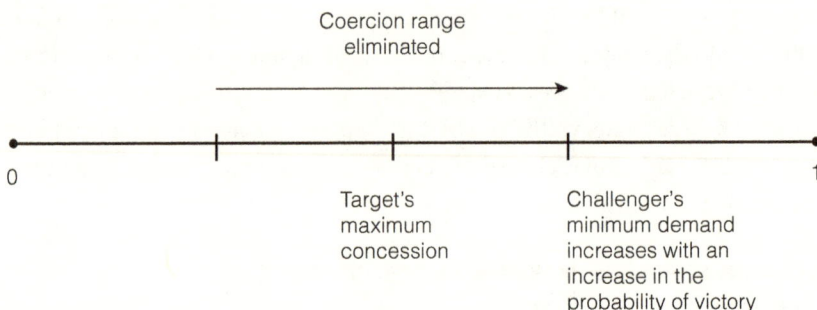

Figure 3.1. No coercion range due to a high probability of victory.

quickly shifted to a brute force strategy. The relatively high expected outcome of a brute force strategy prompted the administration to forego negotiations with the Taliban.

A great power's probability of victory and costs of fighting are not, however, determined solely by the balance of power. Other variables, such as regime type, alliances, geography, climatology, and distance, all factor into the likely outcome of war. Also important is the relative resolve of the challenger and target. In asymmetric interstate crises the great power has the capability to threaten the survival of the weak target state, a fact that elevates the interests at stake for the weak state, which translates into greater resolve.[8] Even with the balance of power in the challenger's favor, if the weak state is resolved to fight until the end, the overall expected outcome of a brute force war may not look all too promising to the great power. In such a case the great power may instead prefer coercion with lesser gains but at a lower cost. For example, the high expected costs of a ground invasion of Kosovo led President Bill Clinton to publicly reject a brute force strategy.

Although a high expected outcome from a brute force war may explain why a great power makes high level coercive demands, it does not explain why a weak state would refuse to make concessions in the face of almost certain defeat. The next section introduces regime change and homeland territory as two sovereignty issues over which a state is likely to run the risks of war.

SOVEREIGNTY AND STATE SURVIVAL

In an asymmetric interstate crisis a great power has the choice of either adopting a coercive strategy and issuing demands or adopting a brute force strategy by seizing objectives through invasion and occupation. Weak states have an incentive to make concessions to avoid a brute force war but are also wary of making concessions that impinge on their sovereignty. Neorealist Ken Waltz defines sovereignty as a state's control over its decision making. Sovereignty, however, does not mean a state can freely choose to do whatever it wants. States face both internal and external constraints. Domestic politics and fiscal budgets determine a state's capabilities, while external pressures from other states and institutions restrict its policy options. Constraint should not be confused with the loss of control over decision making, however, as sovereignty entails a state "decid[ing] for itself how it will cope with its internal and external problems."[9]

Stephen Krasner categorizes three types of sovereignty: domestic sovereignty, which is a state's control of the activities within and transiting its borders; Westphalian sovereignty, which prohibits interference in a state's domestic affairs; and international legal sovereignty, which is an external recognition of a state's authority to negotiate agreements or treaties.[10] Domestic and Westphalian sovereignty are both concerned with a state's control of its homeland territory. At the international level, however, Krasner considers sovereignty as recognition granted by other states and omits Waltz's definition of sovereignty as a state's control over its foreign policy decision making. Such autonomy, however, is a defining characteristic for states that must operate in an anarchic international environment.[11]

With enough coercive pressure, states may be constrained in their policy choices, but they are not likely to concede domestic sovereign control over their homeland territory or international sovereign control over their foreign policy making. Such demands for homeland territory and for regime change do threaten state survival, and these concessions are likely to be resisted.

DOMESTIC SOVEREIGNTY, HOMELAND TERRITORY, AND STATE SURVIVAL

A short time later, the Eastern Hu returned and said: "You have a thousand li of unused land which we want." Mo Tun consulted his advisers. Some said it would be reasonable to cede the land, others that it would not. Mo Tun was enraged and said: "Land is the foundation of the State: How could one give it away?" All those who had advised doing so were beheaded.[12]

—Sun Tzu

States view sovereignty over their homeland as fundamental to their survival. States value control over territory for economic, geostrategic, and nationalist reasons; because it is so important, control over land has long been an issue of international conflict. States require land to be a viable independent actor within the international system, and losing sovereignty over their homeland results in a state's demise, the fate temporarily suffered by Kuwait from August 1990 to March of 1991.

States will resist demands for their homeland territory. Granted, there have been rare occasions when a state has conceded portions of its homeland, as at the outset of World War II, when Czechoslovakia surrendered the Sudetenland in an attempt to appease Nazi Germany, or in 1999 when Serbia relinquished Kosovo to stop NATO bombing. Because the outcomes of international conflict are probabilistic, it should not be surprising to find such exceptional instances where a state has been coerced into giving up a piece of its homeland. In these two examples, however, the states' goal of ensuring its survival necessitated such dire measures. First, Czechoslovakia's decision to relinquish control of the Sudetenland was based on its unpreparedness to fight Germany. Also, its allies England and France abandoned Czechoslovakia to appease Hitler in the hope that such action would satiate German aggression and preserve the remainder of Czechoslovakia. Second, Serbia conceded Kosovo because the fragile Serbian economy, seriously threatened by a continuation of NATO bombing, proved more important than Serbia's economically and geostrategically insignificant territory of Kosovo.[13]

It is not the purpose here to determine just how much territory a state could relinquish and still remain viable. Variation in population, climatology, geology, and a host of other factors make such a calculation unique for each state. To reiterate, states require a homeland, but the claim here is not that all states view any territorial concession as tantamount to state death. The point, rather, is that all states require some homeland territory and the ex ante line in the sand of a nation's borders provides the best observable divide over which a state is not likely to negotiate.[14] Homeland fundamentally differs from the other territory a state controls, as it hosts the majority of a state's population and economic activity and the state's national identity is bound to the territory on which it is born and lives. As such, a target state is likely to perceive a demand for homeland territory as a threat to survival.

INTERNATIONAL SOVEREIGNTY, REGIME CHANGE, AND STATE SURVIVAL

Demands for regime change threaten a weak state's control over its foreign policy making and therefore also its international sovereignty. Krasner recognizes that states control their policy making but argues that they can share sovereignty through bilateral agreements or by joining institutions.[15] The European Union (EU) typifies such an agreement in which states share control over judiciary and monetary policy. Krasner argues that the decision to enter the EU, while initially voluntary, is irrevocable "because its member states are now so intimately linked with one another that withdrawal is not a viable option."[16] In 2013, however, EU secession appeared a viable option for states such as Greece or Great Britain as they considered whether the costs imposed by EU membership were worth the benefits. Regardless of whether these or other countries eventually decide to leave the EU, a state's international sovereignty allows them this option.

In the past, states have exercised their sovereignty by withdrawing from international institutions. Japan withdrew from the League of Nations in 1933, Indonesia from the United Nations for nine months in 1965, and France from NATO's military structure in 1966. It is more difficult, however, to determine whether states retain international sovereignty when they are members of coercive institutions. During the Cold War, Warsaw Pact states were highly constrained in their foreign policy decisions by the threat of Soviet military action. Hungary lost control over its foreign policy when Imre Nagy's decision to withdraw from the Pact in 1956 triggered the Soviet invasion, which installed the compliant Janos Kadar as chairman. Other member states did exercise autonomy, however, as Albania withdrew from the Pact in 1968 in the wake of the Soviet invasion of Czechoslovakia.[17] With the Warsaw Pact it is difficult to distinguish whether states were merely constrained in their foreign policy options or had, in fact, sacrificed their sovereignty. Though an intriguing question, it is beyond the scope of this book on U.S. asymmetric coercion.[18]

In addition to international institutions, Krasner considers how sovereignty might be shared in cases of collapsed and failed states. Intervention may be justified when such dysfunctional regimes prove incapable of governing and when both the failed state and the international community benefit from some form of imposed trusteeship.[19] Shared sovereignty is a method of resuscitating or at least stabilizing a failed state. Examples include the World Bank's long-term involvement in the governance of Chad and the administrative functions performed by UN forces in Kosovo.[20] Although shared sovereignty may be a means of dealing with failing states, this concept is not applicable to viable states with the capability of resisting external influence.

Although states resist giving up autonomy over decision making, they often will entertain constraints to their foreign and domestic policies as the price of admission to international institutions such as the EU, UN, or NATO. In exchange, states expect their sacrifice to be more than made up in economic and/or security gains. Self-constraint, however, should not be misconstrued as an abdication of sovereignty. On only four occasions since World War II have states voluntarily given up international sovereignty: Egypt and Syria in 1958, Tanganyika and Zanzabar in 1964, North and South Yemen in 1990, as well as East and West Germany in 1990.[21] In two of these cases, Syria and South Yemen, the states soon discovered they had made a mistake. Syria regained its autonomy from the Egyptian-controlled United Arab Republic via a military coup in 1961, while South Yemen suffered state death when it was decisively defeated in a civil war in 1994. German reunification is less a case of East Germany conceding sovereignty than of Germany regaining the sovereignty it had lost at the end of World War II. This leaves Zanzibar as the sole case in which a modern state permanently bargained away its international sovereignty.[22]

With regard to the asymmetric coercion model, domestic and international survival concerns place an upper limit on the demands to which a weak state is willing to concede. Although the probability of victory in a brute force war against a great power is likely to be low, conceding to the loss of homeland territory or regime change is even less appealing. By contrast, minor demands to change domestic or foreign policies may constrain a weak state from its preferred policy choices but do not take control over its decision making. Concessions to such demands are less likely to threaten a state's survival, in which case a weak state may be successfully coerced, as in October 1998 when Serbia temporarily amended its policies in Kosovo. Under the threat of NATO air strikes, Serbia began withdrawing its military forces from Kosovo and allowing in international monitors. Though the U.S. coercive strategy convinced Serbia to change its preferred policies, Serbia did not concede its sovereignty. Evidence of Serbian autonomy was displayed within weeks when Serbia reversed its decision and redeployed forces to Kosovo once the Kosovar Liberation Army began to occupy the vacated Serbian military positions.[23]

Figure 3.2 linearly depicts the distinction between coercive demands that constrain and those that take control of a target state's domestic and international sovereignty. At the low end of the spectrum are demands for minor policy changes such as humanitarian policy change in Kosovo. Demands

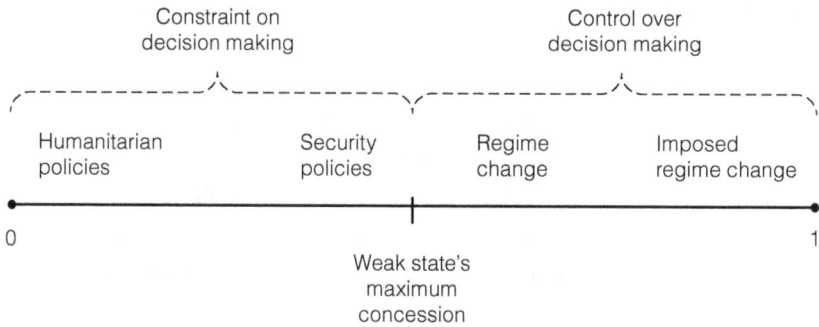

Figure 3.2. Spectrum of demands.

increase when the disputed policies concern more important issues involving national security, as with the U.S. demand in 2002 that Iraq disclose its weapons of mass destruction (WMD) programs. At the upper end of the spectrum are demands to take control over domestic sovereignty by homeland territorial concessions or to give up international sovereignty through regime change. This was the case in March 2003 when Bush demanded that Saddam abdicate power and Iraq allow in foreign occupation forces.

Coercion is likely to fail when, as a result of the high expected outcome from a brute force war, a great power's minimal demand is for the target state to forfeit control over its homeland or its regime. By contrast, the weak state, unwilling to concede to the loss of domestic or international sovereignty, sets its maximum concession below that of homeland territory or regime change. In such a situation, the great power prefers a brute force strategy, even with its risks and costs to lowering its coercive demands. Likewise, the weak state prefers resisting, even though its chances of victory are slight, to conceding to its death.

RELAXING THE UNITARY ACTOR ASSUMPTION
AND TARGET REGIME SURVIVAL

Until now, the asymmetric coercion model has assumed that the challenger and target act as unitary actors.[24] In the real world, a state's decision making is controlled by its ruling regime. In an asymmetric interstate crisis, a weak state's regime is concerned not only with the survival of the state but also with the survival of the regime, which it ensures by remaining in power.[25] At times, the interests of the state and that of the regime may be in conflict, in which case a regime leader may make decisions that lessen the overall security of the state in order to reinforce the regime's hold on power. For example, in 2003 Saddam Hussein adopted the strategy of a ringed defense of Baghdad, ignoring the defense of northern and southern Iraq. He instead emphasized the defense of the Iraqi capital by allowing only his most trusted military forces into the city. His strategy focused primarily on the domestic threats to his regime rather than the external threat to Iraq. Little sense can be made of Iraqi decision making without taking into account Saddam's internal concerns for his regime's security.

Relaxing the unitary actor assumption on the target state allows the asymmetric coercion model to incorporate the survival concerns for both the weak state and its regime. The threat to regime survival can come from two sources: either externally from a powerful challenger's objective of regime change, as with the U.S. demand in 2003 that Saddam leave Iraq, or internally from domestic rivals, a scenario Saddam feared more than the U.S. invasion. Relaxing the unitary actor assumption comes at a cost, however, as it increases the complexity of the asymmetric coercion model. It also dilutes the concept of survival by incorporating not only the survival concerns of the state but also those of the regime and regime leadership. Such a trade-off is warranted, nonetheless, as regime survival is often times at odds with state survival; in such cases as Iraq in 2003, it is difficult to explain the decision making of a targeted state leader without a consideration of regime survival concerns.

The next section considers two different domestic threats to a regime's survival, the threat of rebellion or civil war by domestic opposition groups and the threat of a coup from within the regime.

DOMESTIC THREATS TO REGIME SURVIVAL

In addition to external threats of overthrow, a regime's survival may also be at risk from domestic opposition groups wanting to usurp the government via revolution or civil war. Such groups view a regime's concessions to a great power as a sign of weakness and may seize the opportunity to revolt. A target's ruling regime may therefore resist a challenger's demands out of fear that a concession will reveal the regime as weak and prompt dissident groups to rise up. In such a case, the internal structure of a state is more appropriately viewed as anarchical rather than hierarchical. Steven David coined the term *omni-balancing* to describe the additional internal power dynamic common to weak states.[26] The regime must incorporate into its decision making the likelihood that it will be overthrown by a rival group. In the asymmetric coercion model, the regime must expect these domestic costs for conceding in addition to the international reputation costs the target state already anticipates. The net effect lowers the maximum demand the target state will concede, which, in turn, increases the likelihood coercion fails.

The presence of armed domestic opposition groups within a weak state is an ex ante observable condition. A great power should anticipate a regime's concern over the potential for revolution or civil war. For example, in the early 1990s the rise of domestic terrorist groups in eastern Libya threatened Qaddafi's hold on power. Had he conceded to U.S. demands for Libya to abandon its WMD at that time, he may have placed his regime in jeopardy. By 2003, however, Qaddafi had effectively suppressed these opposition groups and reduced the expected domestic costs for conceding Libya's WMD.

In addition to domestic opposition groups, the regime's leadership may face the internal threat of a coup. Conceding to a great power's demands not only reveals to rival groups that the regime is weak but also signals to its own members that it has suffered a policy failure. With access to only limited information, these members may measure regime performance only on the basis of policy outcomes and are in the position to punish regime leadership for policy failure by removing it from power. Regime leadership should include this risk of a coup into its expected domestic costs for making concessions to a powerful challenger's demands.[27] This interaction between a regime's leadership and its members can be viewed as a principal–agent relationship, in which the leader is the agent charged with carrying out the policy preferences for the principals within the regime.[28] The principals must extract how well the leader adheres to their preferences from the success or failure of the state's policies. If policies fail, the principals may punish the leader by removing him from power. Saddam Hussein, Slobodan Milosevic, and Muammar Qaddafi (see Chapters 4 through 6) understood the threat from a coup, as all three had risen to leadership from within the ranks and seized power when the previous leader appeared to be weak.

Unlike domestic opposition groups, there is no ex ante observable condition to indicate to what degree a regime's leadership is susceptible to a coup. Saddam and Qaddafi took extreme measures to coup proof their regimes, which lowered their risk of being overthrown. However, as their hold on power became more secure they both became increasingly obsessed with their personal protection. The more measures they took, the less safe they felt and the more fearful of a coup they became, no matter how remote the actual chances were.

Previous research suggests that regime type might provide insight into the domestic costs a leader suffers for conceding. The level of domestic costs should vary according to the number of principals within the regime. Democracies could generate high domestic costs because of the relative ease of replacing a leader at the polling booth. The greater transparency within democracies also allows the principals to more easily recognize a leader's foreign policy failures. Autocracies do not have as many principals as do democracies, but the potentially dire consequences for a dictator who loses power makes his domestic costs from a coup significant indeed. It is therefore not obvious whether it is democratic or authoritarian leaders who are more concerned with the loss of face from backing down to a great power's threats. Since World War II, the United States has been involved in crises only against authoritarian states; therefore variation in authoritarian regime type might provide a more useful indicator of a regime leader's susceptibility to a coup. Barbara Geddes developed a categorization for three types of authoritarian regimes: military, single-party, and personalist.[29] Military and single-party regimes typically have more principals involved and are therefore theorized as likely to have higher domestic costs than personalist regimes.[30] Although regime type might provide a useful starting point for the likelihood of a regime leader being removed, empirical work has yet to show that authoritarian regime type is a significant factor in determining asymmetric coercion outcomes.[31]

Though there is no ex ante observable indication as to whether a targeted regime is likely to suffer a coup, a regime's leadership is concerned with saving face before both supporters and opposition. Relaxing the unitary actor assumption on the weak state and expanding target survival to include a regime's domestic survival allows omni-balancing and principal–agent dynamics to provide domestic explanations for coercion failure. For example, the principal–agent relationship helps to understand Milosevic's resistance to U.S. demands for Serbia to concede control of Kosovo. Milosevic feared the domestic political reaction within Serbia had he given up without a fight. In an effort to remain in power he was willing to let Serbia endure seventy-eight days of air strikes.

In regards to domestic regime survival, a final note is warranted on the potential benefits a regime and it leader may also gain for resisting a great power. Weak authoritarian states such as Iraq, Serbia, and Libya live in dangerous neighborhoods. Standing up to a bullying great power and getting a bloodied nose is one way a regime's leader can establish a reputation.[32] By not caving in to external pressure, a leader demonstrates his resolve to opposition groups and regime supporters. Such posturing does not necessarily mean coercion will ultimately fail, though coercive diplomacy likely will. In such a case, the weak target leader needs the great power to employ some of its military force with observable consequences before the leader is willing to concede. In this case, violence serves two purposes. First, as just discussed, withstanding military strikes demonstrates to domestic groups that a regime leader is resolved. Second, military action provides information to both opposition groups and regime supporters as to the true balance of power and resolve of the great power. This demonstrates that further resistance is not only futile but may even threaten the survival of the weak state. In such a situation, even a resolved leader has little choice but to make concessions to preserve the state.

WHY A CHALLENGER CHOOSES A COERCIVE STRATEGY NOT LIKELY TO SUCCEED

The previous section argued that a weak state resists a great power's demands when concession is likely to result in the death of the state or its regime. In theory, such coercive demands for homeland territory or regime change should not materialize due to selection effects. Rationally, a great power anticipating the costs of a doomed coercive strategy should instead select either to accommodate the target or to directly take the objective by brute force. Of course, states do not always behave rationally and, as a result of the biases and uncertainty discussed in Chapter 2, powerful challengers may misperceive and miscalculate and inadvertently set too high a level of coercive demands. For Bosnia in 1992, the European Community and UN International Conference on Former Yugoslavia (ICFY) demanded that the Bosnian Serbs dissolve their newly formed government and concede a significant portion of Bosnian Serb–held territory (see Chapter 5). Though the Republika Srpska had not attained its sovereignty, its leaders acted as any regime would under similar circumstances when survival is at stake by refusing to dissolve the government or concede homeland territory.

Occasionally, great powers miscalculate and implement coercive strategies that fail. There are, however, other situations for which a great power knowingly chooses a coercive strategy that threatens the survival of the target. The next section introduces two explanations for why a great power would rationally adopt a coercive strategy likely to fail, that is, when the costs of adopting a coercive strategy are low and when the diplomatic costs of adopting a brute force strategy are high.

LOW COSTS OF COERCION AND UNCERTAINTY
OVER TARGET RESOLVE

When a great power chooses a brute force strategy, there is often a lag between the decision for war and the deployment of military forces. In such situations, a great power may use coercion as an interim measure while it prepares for war. Because the powerful challenger is already incurring signaling costs by deploying forces or launching preparatory air strikes, the additional costs of a denial strategy are likely to be small.[33] Coercing while preparing for a brute force war makes sense, though, only when there is some hope that the target may concede. The intuition is that it costs the challenger little to make threats because it is already preparing for war. If coercion fails, which is likely, the challenger will have wasted little, but, if coercion succeeds, the challenger avoids the high costs of invasion and occupation.

An example of the low costs of coercion as an interim strategy can be found in the lead-up to the 1991 Gulf War. The United States took six months to build up its troop levels in the Kuwaiti theater of operations to expel the Iraqi Army from Kuwait. In the meantime, the Bush administration adopted a coercive strategy, first leveraging sanctions and then later the threat of air strikes and invasion. Adopting a coercive strategy was not expensive, as the United States would have incurred these expenses in its preparation to invade Kuwait anyway. Had Iraq conceded to UN demands, however, the United States would have avoided the costs of the subsequent invasion.[34] In the end, it cost Bush little to make the demands, and it also reduced the overall costs for war, as the following explanation for the external costs of brute force strategies will elaborate.

EXTERNAL COSTS FOR A CHALLENGER ADOPTING
A BRUTE FORCE STRATEGY

Crises between great power challengers and weak target states transpire within the international environment. As such, a challenger should factor in not only the expected costs and benefits of its interaction with the target state but also the impact such actions will have on third parties. A brute force strategy likely threatens the interests of other states and may generate negative externalities for the great power. Since World War II there has developed an increased expectation that states attempt to resolve their conflicts through negotiation or to bring their disputes before international institutions such as the United Nations. States that abrogate this norm incur diplomatic costs ranging from a general increase in tensions with other states to more serious consequences if a third party is drawn into the conflict. Great power leaders may also be able to dissipate domestic opposition to a war by making an attempt to negotiate a settlement. Attempting to negotiate first may lessen external diplomatic and internal political costs. Therefore, even when coercion is likely to fail, it may still be advantageous for a great power to first go through the UN Security Council before carrying out its preferred brute force strategy. This may not only reduce the diplomatic and political costs of war but may also increase the great power's probability of victory when this gains allies and/or isolates the target state.

The Gulf War is an exceptional case in which the elder Bush administration's use of a coercive strategy, transmitted via Security Council resolutions, not only isolated Iraq and lowered diplomatic costs of the invasion but actually garnered significant resources from allies to pay for much of the war. This contrasts sharply with the junior Bush administration's experience in 2002, when Saddam's conceding to a UN resolution to disclose Iraqi WMD spoiled Bush's attempt to reduce the diplomatic costs for a brute force invasion of Iraq. Bush was subsequently thwarted in his administration's efforts to obtain a further resolution authorizing the removal of Saddam from power. The decision to invade despite opposition by the Security Council was met with dismay by the international community and strained relationships with France and Russia, who had led the campaign within the Security Council to deny authorization.

In both 1991 and 2002 the United States employed coercive strategies in an effort to lower the costs of its preferred brute force strategies of invading Kuwait and Iraqi respectively. In both cases, target concession was even unwelcomed. Although the United States desired coercion success early in the fall of 1990, by the eve of the Gulf War in January of 1991 they no longer sought a peaceful outcome. Having already borne the costs of positioning half a million troops in theater, the elder Bush administration's worst-case scenario would have been an eleventh-hour concession by Saddam that left the Iraqi Army intact.[35] By contrast, in 2003 the junior Bush administration never wanted coercion to succeed but counted on its failure as justification for invading Iraq.[36]

CONCLUSION

This chapter has developed an explanation for why weak states resist great powers. Due to the large asymmetry in power, great powers have an incentive to make large demands of weak states. Yet substantive demands for homeland territory or regime change threaten the survival of a state and/or its regime leading weak target states to resist, even when the probability of victory in a brute force war is small. On occasion, great powers misperceive and miscalculate, choosing coercive strategies that fail. Rationally, a great power should recognize that a weak state will resist concessions that lead to its demise, in which case the great power should either accommodate the target or use brute force to achieve its aims. In other cases, great powers knowingly adopt coercive strategies likely to fail, using coercion as an interim measure in a brute force strategy.

The next chapter employs the asymmetric coercion model and coercion range as a framework to examine two crises pitting the United States against Iraq. The 1991 Gulf War and 2003 Iraq War have been referenced throughout this chapter to illustrate the U.S. choice of coercion over brute force, for weak state's decision to concede or resist, and why the United States would choose coercive strategies expected to fail. Deeper analysis into these two crises provides support to the theoretical claims made in this and the previous chapter.

4 THE UNITED STATES VERSUS IRAQ

The Gulf and Iraq Wars

TWO CONFLICTS PITTING THE UNITED STATES against Iraq from 1990 to 1991 and from 2002 to 2003 are key cases for explaining why U.S. efforts at asymmetric interstate coercion sometimes fail. In both cases, the United States escalated these conflicts into crises. In 1990, it demanded that Iraq withdraw from Kuwait; in 2002, it demanded that Iraq abandon its weapons of mass destruction (WMD). Each time, the United States threatened military action, deployed hundreds of thousands of troops to the region, and then made good on its threats by invading Kuwait in 1991 and Iraq in 2003. Iraq responded to U.S. demands in both cases by making concessions in an effort to forestall the invasions. Though American efforts at coercive diplomacy failed in January 1991 with the launching of the air campaign, by late February coercive air strikes and the impending ground invasion convinced Iraqi President Saddam Hussein to accept a Soviet-brokered peace plan. In contrast, in 2002 U.S. efforts at coercive diplomacy succeeded in getting Iraq to cooperate with UN inspectors and disclose that it no longer possessed WMD.

Ultimately, coercion failed to resolve these crises, as the United States refused to accept Iraqi concessions. In 1991, President George H. W. Bush rejected the conditions of the Soviet peace agreement. Instead, he responded with an ultimatum calling for the immediate withdrawal of Iraqi forces from Kuwait. Saddam refused. In February 2003, Secretary of State Colin Powell argued before the UN Security Council that Iraq had failed to disclose the full extent of its WMD. In both cases, the United States rejected Iraqi concessions and reverted to brute force to achieve its foreign policy goals.

The asymmetric coercion model developed in Chapter 2 provides a framework for analyzing U.S. and Iraqi decision making. The model postulated a sequential logic to asymmetric interstate conflict, whereby the large imbalance in power granted the United States the option of deciding whether or not to accommodate Iraq. When the United States chose to escalate the conflicts into crises, the model identifies coercion and brute force as the two military strategies available. When coercion was chosen, the model highlights the U.S. task of matching threats to demands and communicating the credibility of those threats through the signals of major military deployments and/or air strikes.

The asymmetric coercion model also highlights the constraints placed on Iraqi decision making. As a weaker state, Iraq had little latitude except to decide whether to concede or reject U.S. demands. The coercion range identifies the set of demands resulting in coercion success, that is, concessions for which the United States prefers coercion to brute force and those demands for which Iraq prefers concession to resistance. The great asymmetry in power, however, introduced an incentive for the Americans to make demands on Iraq that proved too costly for Saddam to meet. In February 1991, Bush's ultimatum for Iraq to withdraw immediately from Kuwait aimed at publicly humiliating Saddam by forcing the Iraqi Army to abandon much of its heavy weaponry in Kuwait. In 2003, the younger Bush's demand for Iraq to abandon its WMD aimed at providing justification for the brute force invasion the United States was already planning. Why, then, did American policy makers not disregard coercive strategies it expected to fail and instead gain its objectives through brute force strategies at the outset of these crises?

Implementing a brute force strategy without first attempting to negotiate an agreement abrogates international norms and generates diplomatic costs. These norms expect states to first seek peaceful settlements by way of bilateral negotiations or through international institutions. From August to October 1990 the United States secured ten resolutions through the UN Security Council that condemned Iraq for the invasion and called on it to withdraw from Kuwait. In November 1990, even after Bush had approved plans for a brute force invasion, the United States still sought justification for war per UNSC Resolution 678.[1] Going through the United Nations lowered the anticipated diplomatic costs. It also increased the probability of a U.S. victory, as other states joined the coalition or at least refrained from supporting Iraq. In obtaining the resolution, the United States publicly adopted a coercive strategy meant to achieve an unconditional withdrawal, while garnering authorization for a brute strategy to forcibly liberate Kuwait and achieve its unstated objective of destroying the Republican Guard.

Over a decade later, in November 2002, the UN Security Council passed Resolution 1441, demanding that Iraq disclose its WMD programs and allow enhanced inspections.[2] Unlike Resolution 678, which authorized all means necessary to liberate Kuwait, Resolution 1441 did not authorize the use of force. Subsequently, the George W. Bush administration attempted, but failed, to obtain a second resolution sanctioning Iraqi regime change. A defiant Bush then accepted the diplomatic and political costs of a brute force invasion of Iraq without UN authorization. In both the Gulf War and Iraq War, the United States knowingly adopted coercive strategies likely to fail in an effort to decrease the diplomatic costs and increase the probability of victory for the brute force strategies it preferred.

THE GULF WAR: AUGUST 2, 1990–FEBRUARY 28, 1991

In August 1990, a confrontation occurred between the United States and Iraq, after Iraqi forces invaded Kuwait. Iraq's motivation for the invasion stemmed from its dire financial situation, for which it partially blamed Kuwait, as well as from Saddam's desire to assume a dominant position in the Middle East. Iraq had incurred significant war debts during its eight years of fighting the Iran–Iraq War, with most of the debt held by Saudi Arabia and Kuwait. By 1989, Iraq allocated 22 percent of its government expenditures just to service this debt.[3] Saddam rationalized that Iraq had fought Iran for the benefit of all Arabs; therefore, Saudi and Kuwaiti loans represented contributions to the war effort.[4] Saudi Arabia agreed to financial concessions, but Kuwait refused.[5] In addition, Saddam claimed with some justification that Kuwait was responsible for depressed global oil prices by producing above its OPEC quota. Iraq had then increased its diplomatic pressure on Kuwait to curtail production, but the Kuwaitis ignored Iraqi demands. In late July 1990, Saddam deployed his Republican Guard armored divisions to the Kuwaiti border.[6] Following a brief meeting with the U.S. ambassador, Saddam ordered the invasion of Kuwait on August 2, 1990 (see Map 4.1).[7]

The United States Escalates the Conflict into Crisis

The invasion of Kuwait sparked a long-term conflict between the United States and Iraq. Iraqi actions placed three U.S. interests at risk. First, the invasion destabilized global oil prices that had a direct impact on the American economy.

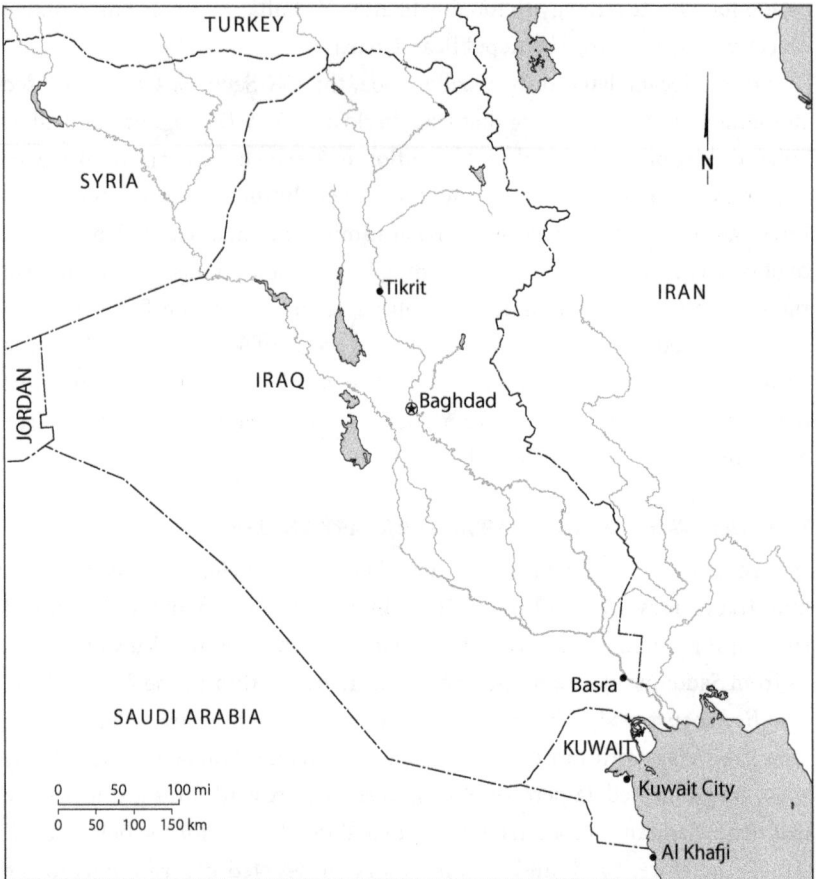

Map 4.1. Iraq.

Saddam now controlled a third of Middle Eastern oil production and approximately 20 percent of the world's proven oil reserves.[8] An already weak U.S.
economy suffered increasingly as crude prices doubled in August 1990.[9] Second, the invasion threatened the ongoing Middle East peace process as Saddam attempted to link the crisis with the Israeli–Palestinian conflict.[10] Third,
American policy makers worried about the long-term impact to international
stability, should Iraqi aggression go unchecked.[11] A failure on the part of the
United States to respond forcibly might undermine Bush's "new world order"
in its infancy.[12]

Though Iraq's action threatened important U.S. interests, Iraq did not have the military capability to threaten the survival of the United States. In contrast, as the sole superpower emerging from the Cold War, America had more than sufficient military might to invade and occupy Iraq. The asymmetric coercion model incorporates these lopsided power dynamics by which the United States had the luxury to choose whether to accommodate Iraq or to escalate the conflict into a crisis by adopting a coercive or brute force strategy.

Iraq may have initiated the conflict by invading Kuwait, but the United States escalated the occupation into a crisis, when Bush ruled out accommodation with these words: "If history teaches us anything, it is that we must resist aggression or it will destroy our freedoms. Appeasement does not work. As was the case in the 1930's, we see in Saddam Hussein an aggressive dictator threatening his neighbors."[13] Bush was concerned about the reputational costs to the United States for appeasing, costs that future crises with Iraq or other states would bear out.

Even though Bush chose not to accommodate Saddam, the United States initially had insufficient forces in the region to force the Iraqi Army out of Kuwait. In addition, a broader concern over international stability led the president to decide to work diplomatically through the UN Security Council.[14] As a result, even if the United States had wished to adopt a brute force strategy at the outset, it did not have the support of three of the permanent members of the Security Council (the Soviet Union, France, and China) to garner a resolution authorizing force.

The United States responded immediately to Saddam's aggression by rushing through the Security Council Resolution 660, which condemned the invasion and issued the compellent demand "that Iraq withdraw immediately and unconditionally all its forces to the positions in which they were located on 1 August 1990."[15] Four days later, the Security Council passed Resolution 661, sanctioning the freezing of Iraqi financial assets and imposing a trade and arms embargo.[16] On August 8, the United States further issued a demand that Iraq not invade Saudi Arabia and deployed its military forces to Saudi Arabia as a signal to make credible its threat to back up this demand. Bush claimed that U.S. forces would "not initiate hostilities, but they [would] defend themselves, the Kingdom of Saudi Arabia, and other friends in the Persian Gulf."[17]

U.S. demands that Iraq withdraw from Kuwait meant that Saddam would have to modify Iraq's foreign policy to give up territory that his army had just seized. Such a concession would not have threatened the Iraqi state's survival, as it would have forced Iraq to cede neither its own territory nor control over its foreign policy. The reason coercive diplomacy failed at this early stage was not that U.S. demands were set so high as to threaten Iraq's survival but that the United States lacked sufficient deployed forces to back up its demands with a credible threat of force.

Within days of the invasion, the United States began deploying its military into the Middle East. The commanding officer of the U.S.-led coalition, U.S. Army General Norman Schwarzkopf, recognized that, until his forces were in place, he did not have the conventional military capability to retaliate against Iraq.[18] As a result, on August 8 he turned to the Pentagon's air staff to develop a coercive punishment strategy. USAF Colonel John Warden, the chief of the Air Force office charged with assessing air campaigns, quickly developed the concept for an intense six-day air operation he christened Instant Thunder.[19] Instead of air strikes designed to punish Iraq, however, this campaign was a coercive denial strategy aimed at attacking Iraq's leadership and command and control targets to produce *strategic paralysis*. As a result, Iraq would have to either concede or be unable to deny the United States from taking its objectives.[20] The air campaign envisioned stealth aircraft delivering precision guided weapons and conventionally armed cruise missiles penetrating Iraqi defenses and striking strategic targets in Baghdad. Meanwhile, tactical aircraft would gain air superiority by rolling back Iraq's integrated air defenses. As more air assets arrived in theater the United States continued to evolve and expand its air campaign plan. Air strikes would not come into play, however, until after the expiration of the UN Security Council deadline of January 15, 1991.

Iraq's Reaction to U.S. Demands and the Threat from Sanctions

U.S. efforts at coercive diplomacy failed as Saddam rejected compellent demands to withdraw from Kuwait and went on to announce the annexation of Kuwait. The effort by the United States to deter an invasion of Saudi Arabia, though a prudent military measure, proved unnecessary because Iraq did not intend further offensive action. Instead, the Iraqi Army prepared defensive positions along the Kuwaiti–Saudi border.[21] To reduce the number of fronts and free up additional troops, Saddam settled the Iran–Iraq stalemate by agreeing to return to the terms of the 1975 treaty.[22] On August 12, in an effort to forestall negotiations, Saddam insisted that any settlement be linked to the Israeli–Palestinian conflict.[23]

The impact of a UN trade embargo on Iraq had an immediate impact on the Iraqi economy, as imports fell by 90 percent and exports by 97 percent. Iraq began rationing in early September.[24] Indeed, Iraq seemed the perfect target for sanctions as it imported nearly 70 percent of its food, and the United States estimated that Iraq had only a two- to three-month supply of wheat, rice, and corn.[25] In addition, Iraqi oil was particularly vulnerable to embargo because its crude flowed primarily through pipelines into Turkey and Saudi Arabia and through tankers navigating the narrow waters of the Persian Gulf.

Economic distress on the part of the Iraqi population, however, did not convince Saddam to withdraw his army from Kuwait.[26] Having ordered chemical weapon attacks against the Iraqi population in the past, Saddam was not moved by civilian suffering and would not change his policies at the first sign of misery. In addition, the sanctions had minimal impact on the Iraqi Army. Having already achieved its objective by seizing Kuwait, the Iraqi Army now set about keeping Kuwait with static defensive positions, which required minimal logistical support. Therefore, though Saddam acknowledged the plight of the Iraqi people, he remained defiant.[27]

On October 5, Soviet President Mikhail Gorbachev sent envoy Yevgeny Primakov to Baghdad to warn Saddam that only a withdrawal would avoid military confrontation.[28] The Iraqi dictator confided to Primakov, "Now that I have given up all the results of the eight-year war with Iran and returned everything to prewar conditions, the Iraqi people will not forgive me for unconditional withdrawal of our troops from Kuwait."[29] Primakov then traveled to Washington on October 19 and presented a gloomy assessment, casting doubts that sanctions alone would coerce Saddam to alter his course.[30]

U.S. Decision to Liberate Kuwait and
Destroy the Republican Guard

In late October, General Colin Powell, the chairman of the joint chiefs of staff, briefed Bush on the need for an additional 200,000 U.S. troops to liberate Kuwait, numbers beyond the 230,000 U.S. and 200,000 international forces already deploying to the Gulf.[31] Powell required a deployment order by the end of October to have these additional forces positioned by mid-January.[32] This forced a decision point for American leaders: They could either continue with economic sanctions and maintain defensive-only forces in Saudi Arabia, or they could ratchet up an offensive military threat. In late October Bush chose the latter, and the subsequent increase in deployed forces both heightened the probability of a U.S. victory as well as lowered the relative costs of fighting a brute force war.[33] Powell declared that this force buildup focused on achieving a decisive military victory.[34]

Bush announced the troop buildup following the November midterm elections.[35] The costs of doubling U.S. deployed force levels signaled the increasing credibility of an offensive military threat. An outcry from Congress, however, diluted the perception of a strong U.S. resolve. Unlike sanctions and the defense of Saudi Arabia, both of which had received broad support, many in Congress disagreed over the offensive military option. Instead, they counseled the White House to show more patience to allow sanctions to work.[36] Polling also indicated that the American public preferred a diplomatic solution over a military approach.[37] To counter this dissent, the president met with bipartisan leadership and top administration officials testified on the necessity of military action before congressional committees.[38]

International support to expel Iraq from Kuwait proved easier to obtain. On November 29, the Security Counsel passed Resolution 678, the twelfth resolution on the Iraq–Kuwait crisis since August 2.[39] The resolution set a deadline of January 15, 1991, for Iraq's unconditional withdrawal and authorized the coalition to take whatever means necessary to enforce Iraq's compliance.[40] Bush, realizing that further diplomatic effort was required to gain congressional support, offered to conduct direct talks with Iraq.[41] This negotiation, however, only fizzled into a brief, unproductive meeting between Iraqi Foreign Minister Tariq Aziz and Secretary of State James Baker on January 9, 1991.[42] Though the effort failed to produce a peaceful settlement, which the Bush administration had not expected to achieve, it did solidify sufficient domestic support for the White House. Three days later, both the House and the Senate authorized the use of force.[43]

Iraqi Reaction to the U.S. Troop Buildup

Iraq responded to the threat of a U.S. invasion of Kuwait in three ways. First, the Iraqis continued to fortify their defenses in Kuwait and mobilized an additional quarter of a million reserve troops.[44] Second, in an ill-fated attempt to garner support, Saddam incrementally released international hostages in a series of high-profile visits by dignitaries and celebrities. Third, Saddam attempted to drive a wedge between the coalition's Arab and Western states by linking any settlement regarding Kuwait to the Israeli–Palestinian conflict. This effort gained some footing within the Arab community following the October 8 confrontation between Palestinians and Israeli defense forces at Haram al-Sharif, an action that left twenty-one Palestinians dead.[45]

A November interview provided insight into Saddam's perception of the crisis. First, he noted that the decision for war was not his to make. He understood the asymmetric dynamics of the conflict and acknowledged that the United States would make the decision. Iraq's only choice was to resist or to accept U.S. demands. Second, Saddam correctly observed that the United States had added the objective of weakening Iraq's military power to its aims. Britain's Prime Minister Margaret Thatcher had expressed this intention to Primakov in October.[46] Though a primary objective of the offensive campaign plan Bush had already approved, he had not made explicit the desire to destroy the Republican Guard, in order to gain and retain domestic and international support.[47] Regardless of whether Saddam was candid, his comments reveal that he understood the asymmetric nature of the conflict and that, in the face of the coalition's growing offensive capability, Iraqi military power was now at stake.

Coercive Diplomacy Failure and the Lead up to the Gulf War

As December wore on, Saddam spoke increasingly pessimistically of the chances for peace; he blamed American intransigence but vowed that Iraq would not withdraw its troops unconditionally.[48] As Iraq prepared for combat, Saddam provided indications of a two-part strategy. First, on December 24, he announced that, in response to any coalition air strikes, Iraq would target Israel.[49] He intended an attack on Israel to elicit an Israeli counterattack that would, in turn, rob the coalition of its Arab support.[50] Second, Saddam predicted that, just as the United States had lost its resolve in Vietnam, it would give up once casualties mounted.[51] Saddam's rhetoric grew ever more bellicose as the January 15 deadline approached. He called on the Arab world for jihad and predicted that Iraq's high moral character would overcome American technological advantage.[52] For Saddam, the key to success lay not in his army's ability to defeat the coalition on the battlefield but in its ability to outlast its enemies by inflicting more casualties than they were willing to endure.[53]

Bush's efforts at coercive diplomacy failed to compel Saddam to withdraw from Kuwait even after his country had endured five months of sanctions and now confronted a war against a superior coalition. In the first stage of the crisis, the United States had not yet deployed sufficient force to threaten to eject the Iraqi Army from Kuwait. By January 1991, however, the credibility of an impending attack had risen dramatically. Yet Saddam still resisted. What explains his continual refusal?

First, misperception and miscalculation over the balance of power and the willingness of the United States to use force figured heavily into Saddam's decision to resist. Iraq's combat experience with Iran, extensive though it was, had not prepared the Iraqi military to face the U.S. military. He underestimated the vulnerability of his army to modern air and ground forces. Saddam also deceived himself by habitually punishing those within his regime who brought him "bad" news. As a result, he misperceived how hollow his forces had become and the degree of technological advantage enjoyed by the coalition forces.[54]

Second, Saddam's dominant and perpetual concern was for his political and personal survival.[55] He had constructed a cult of personality and exhibited the characteristics of Max Weber's charismatic leader. As such, his power rested in his followers' continued belief in his "supernatural" leadership.[56] During the Iran–Iraq War Saddam prevented economic hardship for the Iraqi population by means of foreign loans. The resulting debt was substantial and a driving force behind his decision to invade Kuwait. Yet, the subsequent five and a half months of the international trade embargo only exacerbated Iraqi's economic plight. Unconditional withdrawal from Kuwait could not solve any of these problems and could only further undermine the dictator's authority.[57] Standing up to the United States, on the other hand, would enhance his standing within Iraq and the Arab world and might present him with a political victory, even if it resulted in a military defeat.

A combination of misperception, miscalculation, and political posturing helps explain Saddam's refusal to concede in the face of overwhelming force, but it does not explain the unwillingness of American leaders to compromise. Three factors contributed to their reluctance to reduce demands to settle the crisis once the United States had deployed forces. First, a compromise that left the Iraqi military intact would require the long-term deployment of U.S. forces, lest Kuwait and Saudi Arabia be the victims of future Iraqi aggression. From both a foreign policy and domestic political perspective this represented the Bush administration's nightmare scenario, that is, to have endured the diplomatic and political costs to obtain both UN and congressional approval for an invasion and to have paid the actual deployment costs for half a million personnel, many of whom had already been in place for nearly six months, only to have Saddam accept UN demands at the eleventh hour. On the eve of war, U.S. leaders not only did not want compromise but were relieved when Saddam proved such a "cooperative" enemy by refusing to concede unconditionally.

America's primacy in the nascent post–Cold War world proved to be a second reason for the United States not compromising. Not only had the United States recently emerged as the sole superpower, but it also possessed an abundance of military forces in Europe no longer required to deter a crumbling Soviet Union. These troops could now be deployed to the Persian Gulf. Equally important, neither the Soviet Union nor any other major state had balanced against the United States by aligning with Iraq. As a result, the United States enjoyed an asymmetric advantage in power, which by January translated into a high probability of victory in a war against Iraq at relatively low cost.

Third, U.S. leaders perceived any compromise with Iraq as a threat to broader American interests. Compromise could destabilize the post–Cold War world order by reducing U.S credibility in deterring other states from taking similar aggressive actions.[58] Consequently, they were unwilling to make concessions that would appear to reward Iraq's aggression. In addition, Saddam's insistence on linking any settlement to the broader Middle East peace process only increased the expected costs for the United States in procuring a negotiated agreement and lessened the likelihood of a diplomatic solution convincing Iraq to leave Kuwait.

All told, Iraqi and U.S. calculations over the expected outcome from coercion and brute force war differed so much as to eliminate any coercion range. There were no negotiated settlements to which Iraq would be amenable and that the United States found preferable to war. Coercive diplomacy ended on January 17, 1991, when the coalition commenced air strikes. Although the air strikes signaled American resolve, what remained uncertain was whether they would coerce Saddam prior to a ground invasion.

The Air Campaign

The initial phase of the air campaign prioritized two military objectives, gaining air superiority and attacking Iraq's command and control systems.[59] Coalition air power quickly achieved air superiority; by the end of the first week, the Iraqi Air Force no longer offered resistance.[60] In addition, air strikes wrecked

Iraq's integrated air defense system.[61] The cost to the coalition for achieving air superiority was less than expected, with a loss of seventeen aircraft and their crews (some lost, some captured, some rescued) in the first three days of strikes. By midweek, however, once the United States had restricted aircraft operations to medium altitude, coalition losses decreased dramatically, averaging one aircraft loss per day for the remainder of the war.[62] The coalition could endure such historically low attrition rates indefinitely, as they would not generate the casualties Saddam counted on to undermine coalition resolve.

For strategic strikes, the United States relied primarily on precision weapons dropped by F-111 and F-117 aircraft and strikes by conventionally armed cruise missiles. By the third night, these systems had proven so lethal as to exhaust virtually all Iraqi leadership targets.[63] Unlike previous U.S. strategic bombing campaigns, Instant Thunder was not a punishment strategy aimed at Iraq's population or economy. It was, rather, a denial strategy designed to blind and paralyze the Iraqi government's decision making by interdicting the lines of communication between Saddam and his military.[64]

The initial air strikes failed, however, in paralyzing Iraq as the Iraqis responded to the attacks in three ways. First, they launched ballistic Scud missiles at Israel in an attempt to split the coalition by drawing the Israelis into the conflict.[65] Throughout the war Saddam maintained personal control over Scud missile operations and spoke daily with commanders, to set targeting priorities and make other operational decisions.[66] This provocation placed enormous pressure on Israel's leadership to respond in kind. It required diplomatic pressure by the Bush administration to dissuade Israel from retaliating.[67]

Second, Iraq reacted to strikes on its air force by hiding its aircraft in hardened aircraft shelters, at commercial airports, and even flying some to Iran. Iraq ceded air superiority to the U.S. coalition and instead attempted to preserve what it could of its air defense system. Third, with the air war under way, Iraq made preparations for the ground invasion, which Saddam hoped would produce the high U.S. casualty rates he sought.[68] Iraq initially refused to bow to international pressure to seek a cease-fire and on January 21 publicly rebuffed such an overture made by the Soviets.[69]

By the third week, the air campaign had shifted its focus to the south, the most significant change being the increased emphasis on attacking Iraq's ground forces. Combat losses to coalition air strikes increased the costs for Iraq resisting U.S. demands. The Iraqi Army deployed in the Kuwaiti Theater of Operations (KTO) forty-three divisions, eight of which were elite Republican Guard units. They totaled 336,000 troops, 3,200 tanks, and 2,400 artillery pieces.[70] In the first week, the coalition conducted 938 strikes, primarily with U.S. tactical aircraft and B-52 bombers. These strikes tripled to 2,798 by week two, and attacks on the Republican Guard rose sixteenfold from fifty-three to 805.[71]

Saddam's strategy to draw Israel into the conflict and to inflict a large number of U.S. casualties in a bloody ground war failed. The Bush administration applied diplomatic pressure on Israel not to respond. They deployed Patriot missile batteries to Israel and diverted 7 percent of the tactical strike missions from the KTO into western Iraq to hunt for the Scuds.[72] Though these attacks never destroyed any Scud missile launchers, the air cover reduced the frequency of launches by forcing the Iraqis to adapt their tactics. The combination of U.S. diplomatic pressure, defensive Patriot batteries, and the massive Scud hunt convinced Israel not to retaliate to over three dozen Scud missile attacks, which killed eleven and injured 220 Israelis.[73]

The coalition appeared content with an extended air campaign to reduce Iraqi forces prior to any ground operation.[74] Saddam, however, was ready for the ground campaign to begin and for U.S. casualty rates to rise.[75] U.S. air strikes did not prevent him from traveling from Baghdad to Basra to meet with his commanders on January 27 and then ordering an Iraqi offensive.[76] Soon thereafter, three Iraqi divisions massed along the Kuwait–Saudi border with their lead units occupying the Saudi coastal town of Al Khafji on the evening of January 29. Saddam would later misperceive this operation as a success and regretted that he had not ordered a larger offensive to disrupt the coalition ground forces preparations.[77] The Iraqi forces, however, held Al Khafji for only a day and a half until Saudi and U.S. Marine ground units dislodged them on the morning of January 31. The larger story was the inability of the Iraqis to maneuver and maintain the offensive as coalition aircraft mounted nearly 300 sorties against the exposed enemy armor. Iraqi columns moving south stalled, and the offensive collapsed.[78]

In February, during the third and fourth weeks of the air war, the coalition prioritized attacks on the Iraqi Army. The number of strikes rose to 3,512 in week three and 3,972 in week four.[79] The lethality of strikes increased also as the coalition made two adjustments to its tactics, relaxing altitude restrictions on A-10 attack aircraft and employing F-111s and F-15Es with infrared targeting pods and laser-guided bombs against Iraqi tanks.[80]

Though strategic attacks against leadership targets fell sharply following the first three nights of strikes, in mid-February air planners returned to attack lower-priority command and control bunkers in Baghdad not yet targeted.[81] On February 13 the coalition hit the Al Firdos district bunker in downtown Baghdad.[82] The attack killed several hundred civilians, captured international attention, and forced the United States to cancel strikes on Baghdad for two weeks. It was Iraqi civilian casualties, rather than the U.S. military casualties Saddam had planned for, that caused the United States to change its strategy by restricting air strikes against Baghdad.

Although Iraq stood defiant for the first three weeks of the air campaign, by week four it began to show signs of weakening. On February 11, Iraq announced its willingness to consider a cease-fire.[83] That same day, Soviet envoy Primakov arrived in Baghdad to discuss the conditions for one.[84] He was initially encouraged as Saddam sent a message to Gorbachev, which for the first time dispensed with linking a withdrawal to the Israeli–Palestinian conflict. His hopes were soon frustrated, however, as Saddam agreed only to send his foreign minister, Tariq Aziz, to Moscow for further talks. The peace terms Saddam then issued on February 15, two days after the Al Firdos bombing, were even more demanding than his original August 12 proposal.[85] American leaders dismissed Saddam's proposal but worried that the Soviet efforts might yet succeed in convincing Iraq to agree to the UN resolutions, a development the United States would find difficult to refuse.[86]

By the fifth week of the air campaign (February 15 through 21), coalition air strikes in the KTO reached their apex at over 4,000 sorties flown, accounting for 85 percent of all missions. In preparation for the ground offensive, strikes increasingly attacked Iraq's front line units.[87] Schwarzkopf had set a goal for coalition airpower to cause the attrition of 50 percent of Iraqi armor by February 24, the date for the ground war to start.[88]

Soviet Attempt at Brokering a Peace Agreement

A month into the air campaign Saddam's strategy was not working. The Scud attacks had failed to draw Israel into the war, and the coalition remained intact. Second, the Iraqi offensive at Al Khafji had failed to draw the coalition into an early ground campaign, and the extended air campaign, now focused on the KTO, was having a cumulative impact on Iraqi Army readiness and morale.

As his strategy continued to unravel and losses to his army mounted, Saddam reevaluated the expected costs of a U.S. invasion and the probability of an Iraqi victory. He was now willing to consider a negotiated withdrawal from Kuwait. On February 18 Aziz arrived in Moscow to meet with Gorbachev. The Soviet leader informed him that the Soviet Union would immediately call for a Security Council session to seek a cease-fire, if Iraq declared its readiness to set a time line for a withdrawal of its troops.[89] Aziz returned to Baghdad and then flew back to Moscow on the 21st, prepared to finalize a peace proposal.

By the morning of February 22, Gorbachev and Aziz had reached an agreement on a six-point proposal. In it, "Iraq agree[d] to implement Resolution 660, that is to withdraw all of its troops immediately and unconditionally from Kuwait to the positions they occupied on 1 August 1990."[90] The timetable would have had all Iraqi forces out of Kuwait City within four days and out of Kuwait within twenty-one days.[91] However, Aziz also insisted on the following caveat: "Immediately upon completion of troop withdrawal from Kuwait, the reasons for the passage of the other Security Council resolutions will no longer exist, so said resolutions shall cease to be effective."[92]

Though the proposal did not meet all the conditions set forth in all twelve UNSC resolutions, it did meet the core demand that Iraq remove its forces from Kuwait. Further, the suspension of the resolutions would not take place until after the withdrawal. The Soviet peace proposal also made no reference to linking a withdrawal to a settlement of the Israeli–Palestinian conflict. It was transmitted to Baghdad and, in the early hours of February 23, Saddam accepted the terms.[93]

Conceding to the Soviet peace plan threatened the survival of neither the Iraqi state nor Saddam's regime. The plan called for Iraq to concede none of its homeland territory. As far as Saddam's survival was concerned, the circumstances were now altered. Prior to the air campaign a withdrawal of the Iraqi Army from Kuwait without a fight would have been humiliating.[94] Although Saddam had a firm control of his Ba'ath party and the military, he had feared the domestic repercussions of such a move, particularly from Shi'ite and Kurdish groups. He had confided as much to Primakov in October. This fear, coupled with Saddam's misperception and miscalculation of the probability of victory and the costs of resisting, explain why coercive diplomacy had failed in the lead-up to the January 15, 1991, deadline. After five weeks of air strikes, however, Saddam's calculus had evolved. Now, on the eve of the coalition ground invasion, he understood that further resistance was not likely to result in Iraq keeping Kuwait, while also risking the loss of his army. Yet, concession no longer entailed the same level of humiliation as before because Saddam could, and later did, claim that Iraq had defiantly and valiantly withstood America's air power.[95]

The United States Rejects the Soviet Peace Plan

Having incurred the diplomatic costs of obtaining UN authorization for war, deployment costs for half a million troops, and political costs in gaining congressional approval, Bush had no intention of accepting a negotiated settlement that spared Saddam's Republican Guard. As the Soviet–Iraqi negotiations continued, the coalition military leaders expedited the start date for the ground invasion to deny Saddam the opportunity to accept all the conditions of the UN resolutions.

As negotiations continued in Moscow, Gorbachev kept Bush informed of the progress and let him know that the peace proposal had been formalized and sent to Baghdad.[96] Bush, however, rejected two conditions of the proposal, the extended timetable for withdrawal and the suspension of all twelve of the UNSC resolutions.[97] On February 22, the day before Saddam approved the proposal, Bush preempted the Soviet negotiations by issuing an ultimatum: A coalition cease-fire would be contingent on Iraq's commencement of a withdrawal from Kuwait by noon the next day, February 23, to be completed within forty-eight hours. In addition, Saddam would have to make a statement agreeing to these terms.[98]

With this pronouncement, the United States revealed all its war objectives. Beyond the liberation of Kuwait, the ultimatum also demanded the humiliation of Saddam Hussein. A forty-eight-hour withdrawal would force the Iraqi Army to execute a hasty retreat in which it would have to abandon a large portion of its heavy weaponry. The crippling of Iraq's military power became an objective as soon as the United States was in a position to pursue it.[99] Once Bush had approved military plans and ordered the deployment of forces back in October 1990, his administration's efforts had aimed at securing international and domestic support for just such a brute force strategy. The U.S. position hardened even further as the air campaign progressed. Aircraft losses remained low, while the Iraqi Army proved vulnerable to direct air attack. By February 24, no coercion range existed, as the expected outcome from the brute force invasion exceeded that of the coercive outcome promised by a Soviet brokered peace agreement.

Saddam Rejects Bush's Ultimatum: An Issue of Survival

By February 23, Saddam no longer had any hope that his strategy of attacking Israel and of generating U.S. casualties in the ensuing ground war would sufficiently weaken U.S. resolve to end the war on favorable terms. Saddam was therefore willing to accede to the Soviet peace proposal because it did not threaten the survival of either Iraq or his regime. Bush's ultimatum, however, was a different matter altogether. Iraq's military was vital to Saddam's perception of national security, given the regional threats from Iran and now the coalition. The extent to which the ultimatum threatened Iraqi survival depended on how much military power Iraq expected to lose in a forty-eight-hour withdrawal. Just how much heavy weaponry would the Iraqi Army have abandoned in such a hasty retreat from Kuwait? And what impact would such a retreat have had on the defense of Iraq? Iraqi front line units had over five months to prepare defensive positions against the coalition's offensive. To abandon those positions in two days would have forfeited its heavy arms, especially its towed artillery, along with extensively prepared defensive positions, and would have denied the Iraqi Army sufficient time to regroup behind the Iraq–Kuwait border.

From Table 4-1, an estimate of the reduction in Iraqi military power resulting from the air campaign can be made by comparing Row 1 and Row 2. Row 1

Table 4.1. Iraqi Army weapons estimates in the Kuwaiti theater of operations.

Date	Location	Tanks	APCs	Artillery
January 16, 1991, eve of the air campaign	KTO	3,475	3,080	2,474
February 24, 1991, prior to the U.S. ground invasion	KTO	2,087	2,151	1,322
Counterfactual estimate had Saddam agreed to forty-eight-hour withdrawal	KTO (if all equipment in Kuwait was abandoned)	1,040	1,075	660
March 1, 1991, end of war	KTO In Iraqi Control	842	1,412	279

SOURCE: Anthony Cordesman, Iraq and the War of Sanctions: Conventional Threats and Weapons of Mass Destruction (Westport CT: Praeger, 1999), 68.

NOTE: The January 16 and March 1 numbers were produced through imagery assessment. The February 24 numbers for battle damage assessment were estimated by USCENTCOM. Various battle damage assessments of the damage inflicted by airpower prior to the ground invasion range from 20 to 48 percent. Using the more conservative 20 percent estimate increases the number of weapons Iraq could recover but does not affect the outcome of this analysis. Anthony Cordesman, Iraq and the War of Sanctions: Conventional Threats and Weapons of Mass Destruction (Westport CT: Praeger, 1999), 68.

is the equipment Iraq deployed to the Kuwaiti Theater of Operations (KTO), which included both Kuwait and southern Iraq, and Row 2 is the U.S. estimate of Iraqi weapons that had survived air strikes at the time of the ground invasion. Row 3 then estimates the weapons the Iraqi Army would have retained if it had abandoned all of its equipment in Kuwait (only half of the equipment deployed to the KTO was actually in Kuwait, with the other half kept in reserve along the Iraq–Kuwait border). Row 3 is therefore an upper-end assessment of the maximum number of heavy weapons Iraq could have lost if Saddam had agreed to the forty-eight-hour ultimatum. By contrast, Row 4 lists the number of weapons actually remaining under Iraqi control in the KTO after the cease-fire. A comparison of Row 3 and Row 4 suggests that Iraq would likely have recovered more military equipment by conceding to Bush even if this had meant abandoning all Iraqi heavy weapons in Kuwait. Iraq would have done better militarily had Saddam agreed to the forty-eight-hour ultimatum.

The relevant question is whether the military losses Saddam could have expected for conceding to Bush's ultimatum would have threatened his survival. In August 1990, the Iraqi Army totaled one million troops (fifty-three divisions), 5,500 tanks, 6,000 APCs, and 3,000 artillery pieces.[100] By the time the army regrouped at the end of the Gulf War, it was one-third of its former size, with 350,000 troops (twenty-eight divisions), 2,300 tanks, 2,000 APCs, and 1,000 artillery pieces.[101] Even if Iraq had been able to recover all the weapons that had survived coalition air strikes, this at most would have allowed the Iraqi Army to retain an additional 100,000 men (nine more divisions) in arms.[102]

By conceding to the ultimatum, Iraq would have abandoned much of its heavy weaponry, particularly its artillery, but it still would have recovered more equipment than it lost in the ensuing ground invasion. Overall, however, it would have retained less than half of the army it had at the beginning of the conflict. It would also have withdrawn an exposed army with ineffective air defenses. In such dire circumstances, the Iraqi army would have stood little chance of defending Iraq, had the United States then chosen to invade and overthrow his regime as Saddam expected. At the end of the Gulf War, the U.S. military had in theater over 500,000 ground troops, 2,000 modern battle tanks, and 900 strike aircraft, not counting those of other coalition forces.[103] The United States thus had both a numeric and a qualitative advantage over the Iraqis. It certainly had the deployed military capability to continue on to Baghdad.

Conceding to Bush's ultimatum would have placed the survival of Iraq and of Saddam's regime at risk. Saddam therefore rejected the ultimatum, while reaffirming his support for the Soviet initiative.[104] The United States, in turn, announced that the Soviet proposal "was unacceptable because it did not constitute an unequivocal commitment to an immediate and unconditional withdrawal. Thus, the Iraqi approval of the Soviet proposal is without effect."[105]

The U.S. Ground Invasion: A Brute Force Strategy

In the early hours of February 24, the coalition commenced its ground invasion. U.S. Marines broached the Iraqi forward defenses along the Kuwaiti southeast border and met with little resistance. By nightfall they had raced through a second line of defense.[106] On the morning of the 25th, the Iraqis mounted a counterattack but were driven back within hours.[107] By evening, Kuwaiti resistance was reporting the departure of Iraqi troops from Kuwait City and, in the early hours of February 26, Baghdad Radio announced a general withdrawal of all forces from Kuwait.[108]

The coalition's ground plan called for the Marine advance to be followed by a U.S. Army VII Corps attack on the Republican Guard's exposed western flank, while the XVIII Corps drove deep into Iraq toward Tallil to cut lines of communication with Baghdad. All of this, however, was contingent on the Iraqi Army standing and fighting. Its retreat instead turned into a race to see how much of his army Saddam could salvage before coalition forces could seal off escape routes. At 0800 on February 28, the United States formally declared a cease-fire, and, though the coalition never completed its encirclement, it did capture or destroy 75 percent of Iraqi tanks, 54 percent of its armored personnel carriers, and 89 percent of its artillery pieces located in the KTO.[109] The reduction of Iraq's overall military power, though not total, was significant and greatly degraded Iraq's ability to project power across its borders. The war reduced Iraq's army from 955,000 to 350,000 troops; the Republican Guard had contracted by nearly half, from twelve to seven divisions.[110]

Iraq Defeated

A flurry of diplomatic activity accompanied the rapid succession of events in the ground campaign in an attempt to bring about a cease-fire. After issuing its withdrawal order on February 26, Iraq informed the Soviets, who then called an emergency meeting of the UN Security Council to discuss the matter.[111] In response, the United States announced its conditions for a cease-fire.

It called for Iraq to agree to abide by all twelve of the Security Council Resolutions and for Saddam to publicly renounce the annexation of Kuwait and to accept responsibility for reparations.[112]

The following day, Aziz wrote a letter to the Security Council, in which Iraq agreed to renounce its annexation of Kuwait and pay war reparations but only if it were absolved from the remaining resolutions. When the United States and the Security Council rejected the offer, Aziz produced a second letter to relay that Iraq would abide by all resolutions.[113]

However, events quickly overcame all of these efforts. The Bush administration believed, as it turned out incorrectly, that U.S. forces had already cut off an Iraqi retreat. Prompted further by the expectation of negative press coverage of carnage along the "highway of death" leading from Kuwait City to Basra, the United States moved unilaterally to declare a cease-fire on February 28. Schwarzkopf and Iraqi military leaders then met on March 3 to finalize the peace agreement and bring the war to an end.[114]

Explaining Coercion Failure

Nonrational explanations for war based on misperception and miscalculation as well as the rational explanation based on uncertainty and private information are useful for explaining why coercive diplomacy failed in the lead-up to the January 15 deadline. Saddam miscalculated the probability of victory and the costs for fighting. The air campaign then provided Saddam with "the stinging ice of reality" to update his calculations and reduce uncertainty over the likely outcome of the impending ground invasion.[115] Although these explanations help us to understand why Saddam resisted prior to the air campaign, they do not provide insight as to why Saddam later accepted the Soviet proposal but rejected the Bush forty-eight-hour ultimatum.

Another rationalist explanation for coercion failure would point to commitment problems, as the United States would have been unable to make a credible promise not to make further demands should Iraq have conceded. In this case, however, there is no evidence that Saddam questioned U.S. commitment. In fact, by agreeing to the Soviet proposal, Iraq was counting on a U.S. commitment to a cease-fire. The question of why commitment problems do not appear to have played a factor in the outcome of this crisis may be answered by U.S. actions as the crisis escalated. The Bush administration effectively tied its own hands with three measures that made it more difficult and costly to adopt military objectives beyond that of attacking the Iraqi Army in the KTO. First, the United States engaged the United Nations, in effect making a public promise through UN resolutions with limited demands, which, if subsequently reneged, would have incurred diplomatic costs. Second, the coalition, which included regional states, had a dampening effect on any additional ambitions the United States might have harbored for regime change. Although these countries were willing to join the United States in ejecting Iraqi troops from Kuwait, they would not be party to an invasion and an imposed regime change in Iraq. Third, the Soviet Union played a role in limiting U.S. designs. Although a weakened Gorbachev could not convince Bush to accept the Soviet proposal for a measured Iraqi withdrawal, the Soviet Union still remained a great power. The United States had vital security interests in the continued peaceful decline of the Soviet Union, one of which involved securing the Soviets' vast nuclear arsenal. Such interests would likely have suffered had the United States invaded Iraq.

For the Gulf War, Iraq's survival concerns better explain the outcome of the crisis than do the conventional explanations for war. In terms of the asymmetric coercion model, once the United States had raised the threat to back up its demands, Saddam was willing to concede so long as Iraqi survival was not threatened. He balked, however, at the Bush ultimatum, which threatened his military. Its destruction would have made Iraq vulnerable to invasion.

In addition to threatening the survival of the Iraqi state, a concession to Bush's ultimatum would have also had considerable impact on Saddam and his regime. Notwithstanding the possibility of a foreign/U.S. invasion, such a concession might well have generated a domestic threat. The humiliation of conceding to the Bush ultimatum followed by a panicked retreat from Kuwait would have generated costs from within the Ba'ath party and the Iraqi Republican Guard. Audience costs are the result of a principal–agent problem, whereby those within the regime evaluate the performance of their leader based on outcomes. Should he fail, the regime may punish the leader by removing him from power. Given Saddam's ironfisted control over the government, military, and the Ba'ath party, however, this was not likely to happen. Personalist regimes such as Saddam's are less likely to generate large audience costs because the leader has eliminated those who might pose a threat.[116] In this case, concession to the United States would not likely have generated sufficient audience costs to threaten his survival.

In addition to threats from within the Ba'ath party, a second domestic concern is whether the survival of Saddam's regime was at risk from domestic opposition groups. Concessions by Saddam could have shown his regime to be weak, leaving such groups to assess the regime as vulnerable and then to attempt a revolt.[117] The Shi'ite at the time were in the position to threaten Saddam's regime at the first sign of weakness. The strength of this opposition is evidenced in the March uprisings that followed the Gulf War and nearly toppled the regime.

Gulf War Conclusions

The Gulf War presents an excellent case for examining the dynamics of an asymmetric interstate crisis. It pitted the enormous power of the United States against the weaker power of Iraq. By invading Kuwait, Iraq created a conflict between itself and the United States. As the asymmetric coercion model expected, however, because of its overwhelming power, the United States had the luxury of choosing whether to escalate the conflict into a crisis. The decision not to appease Iraq rested not only on economic interests but also on concern for the reputation costs that the United States, as the hegemon in a new world order, might endure, should Iraq or other states choose to take further aggressive action. Coercive diplomacy initially failed, however, as the impact of sanctions would have taken time, as would the deployment of a significant number of U.S. and coalition forces to the region. By October, Bush had become skeptical as to whether sanctions would ever convince Saddam to retreat from Kuwait. He therefore ordered the doubling of troop strength to the Gulf and approved plans for a brute force strategy aimed at liberating Kuwait and destroying the Iraqi Republican Guard. To gain international and domestic support for war, however, the American diplomatic efforts had to first adopt a coercive strategy. Security Council Resolution 678 demanded an immediate and unconditional Iraqi withdrawal and authorized the U.S.-led coalition to use any means necessary to liberate Kuwait should Iraq fail to comply.

In the final days leading up to the resolution's January 15 deadline, the White House feared coercive diplomacy might actually succeed. The Bush administration had already paid the diplomatic, political, and deployment costs of preparing for war, and its worst-case scenario was for Saddam to now capitulate and withdraw his army from Kuwait. Such an outcome would have freed Kuwait, but it would also have spared the Republican Guard and allowed Saddam to threaten his neighbors in the future. This fear proved unfounded, however, as on the eve of air campaign no coercion range existed for a negotiated outcome that the United States would have preferred to war and that Iraq would have preferred to resistance. This time coercive diplomacy failed because Saddam grossly miscalculated the military balance in the Persian Gulf and vastly overestimated the costs his army could impose on coalition forces in a war.

As the air campaign commenced, the United States initially focused on gaining air superiority and targeted Iraqi leadership in an attempt to paralyze Saddam's control over his military. Although it quickly gained air superiority and its strategic air strikes were highly accurate, efforts to achieve strategic paralysis failed. Saddam was instead able to implement his strategy by launching Scud missiles into Israel in an attempt to split the coalition. When the air campaign continued with little sign of an impending ground invasion, Saddam attempted to increase coalition casualties by jump-starting the ground war with his offensive at Al Khafji. By mid-February, a coercive denial strategy had convinced Saddam to seek a cease-fire, because he finally realized that his strategy was not working. The coalition remained intact; his army was slowly being reduced from the air; and it would soon be threatened with annihilation in the looming ground invasion. As a result, on February 23 Saddam accepted the Soviet peace proposal, a plan that fulfilled the core UN requirement that Iraq withdraw from Kuwait. The key reason Saddam agreed to the Soviet proposal was that doing so did not threaten the survival of his regime or the Iraqi state. Such a peace agreement would have saved the Iraqi Army and not impinged on any of Iraq's homeland territory. Saddam could further save face by claiming that the Iraqi army was returning home victorious after having defiantly resisted the most powerful country on Earth along with its coalition of thirty-three nations.[118]

Though Iraq now preferred the Soviets' proposed coercive outcome to the coalitions' invasion, no coercion range existed for a settlement. U.S. leaders preferred its brute force aims of liberating Kuwait and destroying the Republican Guard to that of only coercing Iraq out of Kuwait. Bush's forty-eight-hour ultimatum effectively elevated his coercive demands to equal that of his war aims. Saddam was now willing to vacate Kuwait to save the army responsible for protecting him and the Iraqi homeland, and he would not concede to a humiliating hasty retreat that would likely destroy it. Such a concession would have made Iraq vulnerable to invasion and revealed Saddam as weak to domestic opposition groups. Saddam therefore preferred the slim chance of an Iraqi draw in a war to that of making concessions that threatened Iraq and his

regime. Coercion was no longer failing because of misperceptions, miscalculation, uncertainty, or commitment problems but because of fear for survival
of the Iraqi state and Saddam's regime.

The asymmetric coercion model and coercion range provide a framework
for understanding the dynamics of this conflict. Survival of the state and regime also provides the most convincing explanation for why coercion ultimately failed, and only the coalition invasion of Kuwait settled the crisis. The
Bush administration chose to maintain a coercive strategy that it did not believe would nor wished to have succeed to lower the international and domestic costs of implementing a brute force strategy. In the next case, Bush's son,
George W. Bush was driven by similar motives in taking diplomatic action
through the United Nations in an ill-fated attempt to justify his administration's invasion of Iraq.

U.S. Foreign Policy toward Iraq: 1991–2002

The Gulf War did not end the conflict between the United States and Iraq, as
Saddam remained in power. The United States adopted a three-pronged strategy to contain Iraq by continuing economic sanctions, deploying UN inspectors to verify Iraq's abandonment of its WMD program, and implementing
a northern safe haven and a southern no-fly zone. Though this strategy was
successful, by the mid-1990s America was growing weary of repetitive crises,
while it perceived its containment efforts as slowly unraveling. The impact of
sanctions significantly declined once Saddam agreed to the UN Oil-for-Food
program in 1996.[119] By 1998, Iraq was exporting billions of dollars worth of oil
each year. As for imports, Saddam controlled to a large extent which countries received lucrative contracts to provide goods to the country in exchange
for the oil, as well as how those goods were distributed within Iraq. December 1998 proved a turning point in U.S. foreign policy, with the decision to
withdraw UN inspectors in preparation for four days of punitive air strikes,
an action named Operation Desert Fox. That decision subsequently left Iraq's
WMD programs unmonitored.[120]

In addition, Iraq grew increasingly defiant in the northern safe area and
the southern no-fly zone, engaging aircraft entering its air space.[121] Convinced
Saddam would never fulfill the obligations of the UN resolutions to which he
had agreed in 1991, the Republican-led Congress passed the Iraq Liberation
Act in 1998, making Iraqi regime change a U.S. mandate.[122] Though regime
change was now its official policy, the United States did not adjust its contain-

ment strategy to eject Saddam from power until the events of September 11 elevated the perceived threat of Iraqi WMD. Following the 2001 invasion of Afghanistan, the United States again turned to Iraq, this time with a resolve to bring about regime change and, in so doing, remove any threat of Iraqi WMD.[123]

The September 11 attacks demonstrated Al Qaeda's capacity to hit targets within the United States and raised the concern over WMD falling into Al Qaeda's hands or those of other international terrorist groups. The George W. Bush administration added Iraq to a short list of state sponsors of terrorism. Though there was little evidence linking Iraq to Al Qaeda, the president justified this action by pointing to the potential threat of WMD being procured from Iraq and used in a terrorist attack against the United States.[124] No longer would the United States regard Iraq's WMD program as a mere regional threat that could be contained by UN inspections and economic sanctions. Saddam's rhetoric did not help matters, when he pronounced the 9/11 attacks a direct result of the "evil policy" of the United States.[125]

THE IRAQ WAR

The September 11, 2001, attacks fundamentally altered America's perception of the threat from WMD. The United States would no longer accommodate states that threatened to supply terrorists with such weapons. Bush's 2002 State of the Union Address escalated the conflict into a crisis when he put Iraq on notice, labeling it a rogue state along with Iran and North Korea in an "axis of evil, arming to threaten the peace of the world." He noted that "by seeking weapons of mass destruction, these regimes pose[d] a grave and growing danger . . . [and] . . . could provide these arms to terrorists, giving them the means to match their hatred."[126] In his June 1, 2002, address at West Point he further articulated his preventive war strategy, assuring his audience that "the war on terror [would] not be won on the defensive" and that the United States would have to "take the battle to the enemy, disrupt its plans, and confront the worst threats before they emerge[d]."[127]

Though both Iran and North Korea had active WMD programs at the time, Bush chose to target Iraq alone. Before the UN General Assembly on September 12, 2002, he demanded that "if the Iraqi regime wishes peace, it will immediately and unconditionally forswear, disclose and remove or destroy all weapons of mass destruction . . . [and] end support of terrorism."[128] As in the lead-up to the Gulf War, to shore up domestic and international support, the White House sought authorization for the use of force from both the U.S. Congress and the UN Security Council. Unlike his father, however, Bush found it more convenient to first secure approval from Congress, where in October both the House and Senate voted by a large bipartisan margin to authorize the use of military force.[129]

By contrast, garnering a Security Council resolution proved more challenging. Neither France nor Russia would condone military action against Iraq. After weeks of diplomatic maneuvering, Secretary of State Colin Powell finally pushed through UNSCR 1441 on November 8, 2002. The resolution held Iraq in "material breach" of previous resolutions and afforded Iraq a final opportunity to comply by disclosing its WMD programs and allowing enhanced inspections. It also warned of serious consequences for noncompliance.[130] To obtain support, Powell had to weaken its terms, whereby Iraq would be considered in material breach of UNSCR 1441 only if found to be falsifying both its weapons declarations *and* (rather than *or*, which the United States preferred) not cooperating fully with UN inspectors. Iraq was to provide unrestricted access for inspectors and declare all aspects of its WMD programs within thirty days.[131]

Though the resolution exerted considerable pressure on Saddam to change his policy on WMD, the Security Council did not threaten invasion or regime change, nor did it authorize a U.S.-led coalition to use force. In his January 2003 State of the Union Address, however, Bush further threatened Iraq's sovereignty by promising to invade, to impose regime change, and to destroy Iraq's WMD, should Saddam not comply with the resolution.[132] The United States signaled the credibility of its threats diplomatically with the passing of UNSCR 144 and militarily with its successful military operations in Afghanistan and with a public release of plans for a rapid buildup of ground forces in the region.[133]

Iraq Concedes to U.S. Demands

Prior to Bush's January 2003 warnings, however, a combination of economic sanctions, UN inspections, and U.S. air strikes had already prompted Saddam to abandon WMD.[134] The Republican Guard had further destroyed all evidence of having disposed of any remaining weapons. Saddam kept this information secret, however, to deter regional enemies. Unfortunately for the fate of his regime, this placed him in the untenable position of having to prove that which he had worked so hard to obscure. A dozen years of deceiving and harassing UN inspectors had left both the United States and the inspectors ill disposed to lend credence to any statement coming out of Iraq.

Because Bush had labeled Iraq a founding member of the axis of evil in January 2002, Iraq had been in discussions with the UN Secretary General over the potential return of inspectors. Though Iraq had consistently refuted U.S. accusations concerning its WMD programs, the gravity of Bush's September 12, 2002, UN address was not lost on the Iraqis. Within days, Iraq agreed to the return of inspectors "to remove any doubts that Iraq still possesse[d] weapons of mass destruction."[135] Though denying having WMD and opposed to UN resolution 1441, once it had passed, Iraq reluctantly agreed to abide by it, allowing UN inspectors into the country and providing a 12,000-page declaration on December 7, 2002.[136]

Iraq's WMD capabilities had eroded over the previous decade as a result of sanctions preventing the importation of unconventional weapons equipment and technologies, by a period of cooperation with UN weapons inspections in the early 1990s, which destroyed much of Iraqi stockpiles, and by air strikes that destroyed known WMD facilities.[137] As a result, Saddam either could not or chose not to reconstitute Iraq's WMD program.[138] Instead, he adopted a deterrent strategy of ambiguity over the status of WMD.[139] The impetus behind this unconventional weapons strategy could, in part, be traced to the destruction of much of Iraq's conventional military power during the Gulf War. Iraq's 350,000 soldiers were now a mere shadow of the nearly million man army it had previously fielded prior to the Gulf War.[140]

With Iraq now anticipating the return of UN inspectors, Saddam ordered all WMD sites to be scrubbed of any remaining documents or traces of chemical, biological, or nuclear weapons.[141] Ironically, it would be the actions of the Iraqi military in sanitizing these locations that Secretary of State Colin Powell would later offer as "proof" of Iraqi deception.[142]

Had the United States accepted these Iraqi concessions to UNSCR 1441, this crisis would have terminated and now be remembered as a successful case of coercive diplomacy. It also would have satisfied the equilibrium conditions of the asymmetric coercion model, which expects coercion to succeed. Rather than accommodate Iraq, however, the United States chose to escalate the conflict into a crisis by adopting a coercive strategy. U.S. leaders demanded that Iraq abandon its WMD. Because Iraq no longer possessed any WMD, the United States was, in effect, demanding Saddam make the moderate policy change of admitting to that fact.[143] It backed this demand with the threat of force, made credible by the costly signals generated by diplomatic and military action. Given the credibility of U.S. threats and, as argued next, the fact that Iraqi concession to these demands did not threaten the survival of Iraq or his regime, Saddam could be coerced into abandoning the pretense that Iraq still possessed WMD.

Iraq was, in fact, coerced even though it did not find the U.S. threat of regime change credible. The United States had signaled the credibility of its military threat through its successful operations in Afghanistan, its ongoing aggressive air operations in the southern no-fly zone and northern safe area, and in its additional deployment of forces to the Persian Gulf. Though the threat of air strikes was credible, Saddam did not believe the United States would incur the high costs of invading and occupying Iraq. Because Iraq no longer possessed WMD, he incorrectly expected that the disclosure of this fact and his cooperation with UN inspectors could forestall an invasion. He likely thought that if the United States were to attack, it would limit its actions to air strikes as it had in Desert Fox in 1998 or occupy peripheral territory such as Basra.[144] In the end, the threat of restricted air strikes alone, however, was sufficient to convince Saddam to admit to no longer having WMD.

The asymmetric coercion model expects coercion to succeed in this case, as the United States should only adopt a coercive strategy and issue compellent demands to which Iraq would be willing to concede. Whereas Chapter 3 introduced state and regime survival as reasons for why a weak target state might resist, here the demand for Saddam to abandon his policy of ambiguity over WMD threatened neither Iraq's survival nor that of Saddam's regime. This is not to suggest that he was not concerned over his personal safety or the future of Iraq. In fact, he continually adopted measures to protect every aspect of his regime. His WMD policy was an effort to deter aggression by Iran, his reorganization of homeland defense by regions represented an effort to prevent revolution, and his internal security network was to protect him against coups. Even without WMD, the Iraqi Army was strong enough to deter Iran, and Saddam's elaborate security measures were more than sufficient to protect him and his regime from internal and domestic threats.

The most credible regional threat to the Iraqi state was that posed by Iran.[145] Having employed chemical weapons against Iran in the past, Iraq's threat of WMD likely had an impact on Iranian calculations for future military action. Still, given the relative weakness of Iranian ground forces, the Iraqi Army was well equipped to defend itself against an Iranian invasion.

The Iraqi Army numbered 350,000 troops and 2,300 tanks compared to Iran's 305,000 troops and 700 tanks. An Iranian invasion would likely have bogged down in the southeast, as had twice happened in the Iran–Iraq War. Iran thus did not pose a threat to Baghdad and, by extension, to the survival of Iraq or Saddam, personally. The other credible external threat to Iraq was that posed by the United States. Saddam's ambiguity over WMD, however, did not deter the United States. These unconventional weapons had not deterred it in 1991, when Iraq actually possessed them; since the terrorist attacks of September 11, 2001, they served more as a catalyst than a deterrent to war.

Besides its impact on Iran, the threat of Iraqi WMD also had a potential deterrent effect on the Shi'a and the Kurds. Concessions by Saddam would not only alleviate the threat of WMD to these domestic opposition groups but would also signal that his regime was weak. Saddam, however, had successfully put down the March uprising after the Gulf War and had in the ensuing years taken measures to suppress domestic opposition to prevent another revolt. In 1995 Saddam founded the Fedayeen, his loyal paramilitary organization, which gave local Ba'ath parties additional forces to keep towns and villages in line.[146] In the wake of the Desert Fox air strikes of 1998, Saddam also split the administration of Iraq into four regions, delegating control of the military forces in each district to a trusted Ba'ath politician.[147] Though this would later hinder the Iraqi military in defending against the American invasion, Saddam initiated this reorganization to retain better domestic control over Iraq and prevent the rise of rivals from within the army.

Saddam had taken effective measures to reduce the threat of opposition within his borders, but what about the threat from within his regime? As the leader of an authoritarian, personalist regime, Saddam should have expected lower audience costs for foreign policy failure than could leaders of democracies or of military or single party regimes.[148] In addition, Saddam had in recent years expended enormous resources on his elaborate internal security network to monitor not only the Iraqi population but also those within his own military and security apparatus.[149] To further protect against a military coup, he forbade the army and the Republican Guards, save his loyal Special Guard, from entering Baghdad. He had insulated himself from a coup to the extent that concession to U.S. demands did not threaten his ouster from within.

Revealing that Iraq no longer had WMD did not threaten the survival of the Iraqi state, of the Ba'athist regime, or of Saddam's place within that regime. As such, Iraq could be coerced. By contrast, conventional explanations for war caused by misperception and miscalculation or by uncertainty and private information do not explain why Iraq conceded, but they still provide insights into why the United States escalated the conflict into a crisis in the first place. Saddam had an incentive to keep information private over Iraq's WMD to deter Iran. He intentionally introduced uncertainty; because of this, the Bush administration misperceived the threat of Iraqi WMD to be a vital U.S. national security interest. Saddam, however, did not miscalculate by conceding to UNSC Resolution 1441, as he correctly perceived the credible threat of a U.S. attack, had he resisted. Though the United States still invaded, Iraqi efforts did force U.S. leaders to conduct a preventive war without a UN mandate and, in the process, reduce the international support for American action, as well as limit the size of the coalition.

Finally, in regard to commitment problems as another explanation for war, as in the Gulf War, Iraq did not resist out of concern the United States would later raise its demands. Saddam, in fact, quickly conceded to U.S. demands in 2002, after which the United States still went on to make even greater demands for regime change. Given the logic of commitment problems, Iraq should not have conceded and shown itself to be militarily vulnerable as well as weakly resolved. In sum, while other explanations for war do not explain why Iraq conceded to change its policy over WMD, the survival explanation does explain why Iraq could be coerced. Convincing the United States that it had, in fact, abandoned WMD, however, represented a much more difficult proposition.

The United States Rejects Iraqi Concessions

As in February 1991, when Saddam Hussein attempted to concede to the Soviet peace proposal, in 2003 the United States once again refused to accept "yes" for an answer. The Bush II administration used Iraqi WMD as its justification for a long-held foreign policy objective of removing Saddam from power.[150] Deputy Secretary of Defense Paul Wolfowitz even suggested an invasion of Iraq as an initial response to the September 11, 2001, attacks, despite having no evidence linking Iraq to the attacks.[151] This objective of regime change explains the administration's refusal to accept Iraq's WMD declaration and its eagerness to declare Iraq to be in material breach of the UN resolution.

Though the United States had demanded Iraq abandon its WMD, this did not mean that the Bush administration intended to accept a concession. Within a week of his speech before the United Nations, President Bush began making the case against the credibility of any claims or actions by Saddam:

"He deceives. He delays. He denies. And the United States and, I'm convinced, the world community, aren't going to fall for that kind of rhetoric by him again."[152] Bush was unwilling to accept that, after twelve years of defiance, Saddam would finally relent and could now be trusted. He remained convinced Iraq had WMD, and there was nothing Saddam could do to satisfy him. Bush placed Saddam in a position of "Catch-22": if he continued to deny Iraq had any WMD, the White House would label him a liar, and, if he produced evidence of WMD, the White House would call him a cheat.[153] Either way, the Bush administration would have their justification for war.

Vice President Dick Cheney elaborated on this skepticism regarding WMD inspections. In his view, Iraqi cooperation with UN inspectors provided no assurances of Iraqi compliance; "On the contrary, there is a great danger that it would provide false comfort that Saddam was somehow back in his box. Meanwhile, he would continue to plot."[154] Saddam's previous decision to destroy all evidence of WMD would later make it impossible for Iraq to account for the whereabouts of its stockpiles. As a result, Iraqi officials were unable to dispute U.S. claims that these supplies were, in fact, hidden.

In response to Iraq's December 2002 declaration that it no longer had WMD, the United States immediately countered that the report had significant omissions and contained little substantive information on Iraq's program since the departure of UN inspectors in 1998.[155] UNMOVIC (UN Monitoring Verification and Inspection Commission) Chairman Hans Blix and IAEA (International Atomic Energy Agency) Chairman Mohamed ElBaradei provided a similar assessment in their report to the Security Council on December 19, 2002. Although they did indicate that Iraq was cooperating in the process by allowing inspections, they felt more cooperation was required in terms of the "uncovering of evidence to exonerate themselves that they [were] clean from weapons of mass destruction."[156] Secretary of State Colin Powell subsequently declared Iraq in material breach of Resolution 1441. He cited not only the large omissions and gaps in the report but also the disturbing, though as it turned out, truthful "Iraqi declaration [that] denies the existence of any prohibited weapons programs at all."[157]

In hindsight, the United States obviously miscalculated that Iraq still had WMD.[158] Although the Bush administration intended to use WMD as justification for regime change, it was also convinced Iraq had hidden or reconstituted its WMD. The White House was surprised when the military teams sent into Iraq to locate and document the extent of Iraq's programs came up empty handed.[159] This miscalculation elevated the disarmament of Iraq and the removal of Saddam from power to the level of a vital national security interest. As a result, the administration was willing to endure the high costs of a brute force invasion and occupation.

On January 27, 2003, Hans Blix provided his sixty-day assessment of UN inspections, indicating that little had changed since December, that is, that Iraq was cooperating with the inspection process but was still not revealing its WMD.[160] Bush, in his second State of the Union Address on the following evening, asserted that, if Saddam did not fully disarm, the United States would lead a coalition to disarm him.[161] On February 5, Secretary Powell, before the UN Security Council, infamously presented additional evidence from U.S. intelligence sources on Iraqi WMD and ties to terrorist organizations.[162] The following day, Bush ordered 15,000 more troops to join the 100,000 American personnel already in place in the Persian Gulf.[163]

Iraq Denies the United States Its Casus Belli

In response to U.S. accusations that Iraq was in material breach of Resolution 1441, Saddam's new strategy was to cooperate fully with inspectors to prevent the justification for an attack.[164] This was initially difficult for Saddam to do, as a dozen years of deceiving UN inspectors had inculcated the Iraqi government with a culture of obfuscation. Following Hans Blix and Mohamed ElBaradei's initial reports of anecdotal examples of noncooperation, however, Saddam redressed this problem by making it clear to subordinates under the threat of severe punishment that Iraq would fully cooperate with inspectors.[165] A second problem Saddam could never overcome, however, was the thoroughness of Iraq's cleanup, a measure that left him with no proof that the tons of VX nerve agents and Anthrax unaccounted for were not hidden away as the Americans claimed.[166] Still, Iraq cooperated with inspectors by providing them heretofore unprecedented access to the Iraqi military and Saddam's presidential palaces. Iraq even destroyed the seventy-six Al Samud II medium-range missiles it had only recently produced, when U.S. leaders argued the missile's operational range breached existing UN resolutions.[167] By spring 2003, though Saddam had come to believe that the United States could well attack, Iraq continued to cooperate with inspectors to provide additional support for France and Russia in the hopes that their opposition could still halt the United States from its intentions.[168]

U.S. Diplomatic Preparations for a Brute Force War

Powell's presentation before the Security Council on February 5, 2003, was not sufficient to convince France and Russia, who favored the continuation of inspections as an alternative to war.[169] Though the United States was predisposed to going forward with military action on the basis of UNSCR 1441, British Prime Minister Tony Blair desired further authorization to use force from the Security Council to quell his Parliament's concerns over the legality of an invasion. Not wishing to go it entirely alone and as a favor to Blair, Bush pressed forward for the second resolution.[170] France in particular actively campaigned against this proposal. Doubting France would veto, however, Bush announced on March 6 that the United States, Britain, and Spain would jointly submit a resolution to the Security Council "stating that Iraq ha[d] failed to meet the requirements of Resolution 1441."[171] Britain introduced the draft resolution the next day, setting a March 17 deadline for Iraq to disarm.[172]

However, the United States and Britain failed to garner the requisite votes to pass the resolution, and France, Russia, and China stated their intentions to veto, if necessary. On March 16, British Prime Minister Blair, U.S. President Bush, and Spanish Prime Minister Jose Maria Aznar met briefly in the Azores, issuing the following joint statement: "If Saddam refuses even now to cooperate fully with the United Nations, he brings on himself the serious consequences foreseen in UNSCR 1441 and previous resolutions."[173] The next morning, the United States withdrew the draft resolution from consideration, and that very evening, Bush addressed the nation and delivered an ultimatum: "Saddam Hussein and his sons must leave Iraq within 48 hours. Their refusal to do so will result in military conflict commenced at a time of our choosing."[174]

In December, when Powell had declared Iraq in material breach of UNSCR 1441, the administration had abandoned its pretense of a coercive strategy and instead pushed forward with its preferred brute force strategy to depose Saddam.[175] The United States did not match its coercive demands to effect that regime change, as Bush did not believe Saddam would ever voluntarily abdicate power.[176] In fact, it was the threat of forcing regime change that the White House had levied against its demand for Iraq to disarm. This confirms that U.S. leaders did not consider regime change to be within the range of possible coercive outcomes but an objective that could only be achieved through brute force.

The administration did not intend for nor believe that the ultimatum could coerce Saddam from power. In fact, National Security Advisor Condoleezza Rice favored an alternative draft of Bush's ultimatum speech, in which the president would simply announce pending military action.[177] It did not make sense to her to declare an ultimatum, if the United States intended to invade Iraq regardless of Saddam's response. Prior to the end of the forty-eight-hour deadline, in fact, the White House went on to make public its intention to enter Iraq, even if Saddam were to abdicate power and accept exile.[178] In later developments, although Bush officials hinted that the United States might indeed accept an offer of exile for Saddam, Bush would make no guarantees for the safety of Saddam or his family, should they follow through with the offer.[179] Saddam confirmed the validity of Bush's assumption that the United States could not coerce him into either regime change or allowing in foreign troops.[180] He remained defiant, rebuffing any suggestion that he go into exile.[181] Instead he prepared for the air war he expected would come.

Iraq's Military Preparations

On the eve of war, Iraq had an estimated 389,000 active duty personnel, of whom 350,000 were Iraqi Army, a force with an inventory of 2,600 tanks and 2,400 artillery pieces.[182] A subset of the Iraqi Army, the Republican Guard, was comprised of 70,000 personnel divided into three elite groups: The Special Guard in Baghdad provided for Saddam's personal protection, while the Northern, or I Corps, further defended Baghdad and Saddam's hometown of Tikrit from the Kurds, Turkey, and Iran. The Southern, or II Corps, employed its armored divisions to suppress the Shi'as and defend against an Iranian or U.S. attack.[183] Saddam could also count on his Fedayeen. Some 40,000 strong, though not professional soldiers, these men were loyal, had training in small arms, and were capable of conducting fanatical attacks, as the United States would later discover.[184] They reported outside of the army command structure directly to Saddam's sons.[185]

In preparation for the U.S. attack, the Republican Guard withdrew toward Baghdad. Rather than have its soldiers prepare entrenched defensive positions, however, Saddam dispersed his forces for the extended air campaign he felt would come. It was evident he did not anticipate a ground invasion all the way to Baghdad, as he did not take such defensive measures as flooding the Euphrates, setting fire to oil fields, or defending key strategic points along the road to Baghdad. Ignoring military advice, Saddam deployed the army in a series of concentric circles around Baghdad, with his most loyal forces in the innermost rings.[186]

Saddam clearly miscalculated in not anticipating the U.S. invasion, preparing instead against domestic threats to his personal survival and that of his

THE GULF AND IRAQ WARS 83

regime. He had witnessed the fall of the Taliban in Afghanistan to the ragtag Northern Alliance Army supported by U.S. air power. He had also experienced the near toppling of his regime by Shi'a and Kurdish opposition in the March uprisings of 1991. From this point of view, Saddam's decisions made sense. Why endure the costs of extreme defensive measures or concentrate his army in positions that U.S. ground forces would never threaten? He instead set about insulating his regime and positioning his remaining forces to ride out air strikes, as he had done before.

U.S. Military Planning, Deployment, and Combat Operations

Planning for an invasion of Iraq began in the midst of the Afghanistan War. On November 21, 2001, Bush informed Secretary of Defense Donald Rumsfeld that he wanted to "know what the options [were]."[187] The existing contingency Operation Plan 1003-98 was similar to planning for the Gulf War in that it included a six-month buildup of 400,000 troops prior to invasion.[188] Rumsfeld, however, pushed for a lighter and quicker response. By February 2002, General Tommy Franks, Commander-in-Chief, U.S. Central Command, had taken lessons from Afghanistan and pared down the forces required to commence the attack to 160,000. Deployments would then continue until a total of 250,000 troops would be available for occupation at the end of combat operations.[189]

Saddam and the world were well aware of the U.S. military plans, as senior defense officials leaked the war plan just days after the Security Council's vote on UNSCR 1441. In late November, Franks submitted a request for deployment orders. The number of initial forces had now been ratcheted down to 128,000 to be in the region by mid-February 2003. A total of 200,000 were to be in place by the commencement of ground operations.[190] By March 20, the United States had 115,000 American ground troops in place alongside 26,000 British soldiers and marines.[191] Coalition forces in the entire region totaled 250,000, including 735 fixed-wing combat aircraft.[192]

Unlike the Gulf War, for which a lengthy air campaign preceded ground operations, the U.S. planned a near-simultaneous attack dubbed "Shock and Awe." The ground invasion was to be a two-pronged push to Baghdad, with the Marines approaching from east of the Euphrates and the Army maneuvering from the west. CENTCOM abandoned the plan for a northern attack in March when the Turkish parliament voted to deny the United States permission to deploy its 4th Infantry Division from Turkish territory.[193] Strategic air strikes against leadership targets and the Republican Guard had not been planned in advance of the ground invasion. Tactical air strikes, however, had begun by July 2002, as CENTCOM incrementally degraded Iraq's air defenses and command and control network in the southern no-fly zone in preparation for air support for the ground invasion.[194]

Bush ordered the first attacks to commence in the early hours of March 20, just after the deadline for Saddam and his sons to be out of the country had expired.[195] Initial air and missile strikes on Baghdad failed to topple Saddam's regime. By April 9, however, U.S. ground troops had entered central Baghdad.[196] On May 1, 2003, aboard USS *Abraham Lincoln,* the president announced the end of major combat operations in Iraq.[197]

Miscalculating the Costs of the Iraq War

The United States correctly assessed the high probability of victory of a brute force invasion of Iraq and the minimal costs of fighting a conventional campaign. Where Bush grossly miscalculated was in the expected costs for occupation. The most liberal estimate prior to the war had placed the total cost for the war as high as $100 billion. The administration disputed these numbers. It expected that Iraqi oil revenues would pay much of the cost of occupation and reconstruction.[198] Tragically, the actual costs, in terms of U.S. blood and treasure, proved much higher. As of 2011 a conservative estimate indicated the cost at over $832 billion, of which only $53 billion was incurred in 2003 during conventional combat operations and initial occupation.[199] U.S. casualties also proved much higher following President Bush's May 1, 2003, victory speech. Although only 139 personnel were lost during the invasion, an additional 4,347 deaths have occurred in ensuring years (as of April 1, 2014).[200]

The actual costs of war rarely turn out to be the same as the ex ante expected costs, on which the decision for war rests. Though the Bush administration expected occupation costs to be quite low, some in the U.S. Army leadership were skeptical. U.S. Army Chief of Staff General Eric Shinseki was sharply criticized by Rumsfeld after he testified before Congress that he believed it would take several hundred thousand troops to occupy Iraq.[201] Because the administration believed U.S. troops would be welcomed as liberators, it did not authorize the additional deployment of troops required for a phase IV occupation. Had it done so, the expected costs of occupation would have been substantial but still likely lower than the actual costs incurred once the insurgency erupted. Likewise, if the administration had not made the critical error of ordering the disbandment of the Iraqi Army and the de-Ba'athification of the Iraqi civil service, the costs of occupation would likely have been lower.[202] If, in its calculations, the administration had estimated a substantially higher cost for occupation, this could have altered its decision for a brute force war. This might, in turn, have produced a viable coercion range, wherein the United States would have been willing to accept a coercive outcome short of regime change, which Saddam might have reluctantly accepted.

Though miscalculating the costs of occupation explains why the United States preferred war, it does not explain why its leaders first tried to coerce. According to the asymmetric coercion model, coercion should not have failed because it should never have been tried. Rationally the United States should not have adopted a coercive strategy with a lower expected outcome than that of a brute force strategy. The administration, however, understood that it could not achieve its primary foreign policy objectives of regime change and disarming Iraq by coercion alone. Iraq would certainly resist such demands as threatening to the sovereignty of the state and the survival of the regime.

Still, the United States pressured the UN Security Council to obtain Resolution 1441 with its limited coercive demand for Iraqi cooperation and the disclosure of its WMD. The administration, did not, however, expect Iraq to comply. In fact, it counted on coercion failure to justify the invasion it had already planned. Instead of making coercive demands of regime change and occupation, Bush instead threatened them as punishment for the Iraqi non-compliance he expected. The United States did not go to the Security Council to coerce Iraq but rather to obtain a casus belli for a brute force invasion. An authorization to use force would lower the diplomatic and political costs for the invasion, as it had in 1991 with the invasion of Kuwait. Even though the administration had little regard for the UN, it still recognized the value of first going through the Security Council to decrease the overall costs of war. Unlike his father, however, George W. Bush failed to obtain his UN justification for war. In the end, due to his misperception of the threat of Iraqi WMD and miscalculation of the costs of occupation, he accepted the censure by the international community for abrogating international norms of unsanctioned aggression.

Iraq War Conclusions

As with the Gulf War, the Iraq War presents an excellent case to examine an asymmetric interstate crisis, this time with a powerful United States against a much weaker Iraq. Following the Gulf War, conflict continued between the United States and Iraq, primarily over WMD. By 1998 U.S. leaders were convinced Saddam was the problem and that Iraq would never give up its WMD as long as he remained in power. The United States did not, however, have the resolve to take military action until after September 11, 2001. As the asymmetric coercion model expected, because of its overwhelming power advantage, the United States had the option of escalating the conflict into another crisis. After U.S. military forces toppled the Taliban in Afghanistan, Bush chose to no longer accommodate Saddam's regime. He did not, however, demand Iraqi regime change. Instead, in his September 12, 2002 speech before the United Nations, he demanded that Iraq abandon its WMD. Coercive diplomacy initially succeeded, as the costly signals of U.S. air operations in Afghanistan and the deployment of additional troops to the Gulf made credible the threat of U.S. military action and convinced Saddam to abide by UNSC Resolution 1441.

Incredibly, at least to the other members of the Security Council save Britain, the United States refused to recognize Iraq's concessions. The administration had not intended Resolution 1441 to succeed. It had believed the resolution would fail and thereby justify an invasion. By first going through the UN Security Council, the Bush administration expected to lower the diplomatic and political costs for its brute force strategy. Iraq acquiesced, however, and France and Russia judged Iraq in compliance with UNSC resolution 1441, thus denying a second resolution authorizing force. Nonetheless the United States chose to invade. The administration calculated the expected outcome of a brute force strategy to topple Saddam's regime and occupy Iraq to offset the diplomatic costs of going to war without the Security Council's blessing.

The administration, however, made two major blunders in this crisis: First, it misperceived that Iraq still had WMD; second, it miscalculated that the costs of occupying Iraq as incidental. The misperception of the threat of Iraqi WMD increased the U.S. valuation of the core objectives of regime change and disarmament. The miscalculation over the costs of occupation increased the cost of the brute force strategy that eliminated the coercion range, that is, the demands to which Iraq would have been willing to concede and for which the U.S. would have preferred coercion to brute force war. Saddam was willing to abandon a policy of ambiguity over WMD, as doing so did not threaten the survival of Iraq or his regime, but he would not be coerced into regime change and a foreign occupation. By the time of Powell's February presentation before the Security Council, the White House had abandoned the pretense of coercing Iraq. Rather than continue with the coercive demands of Resolution 1441, Powell's objective was to obtain a second resolution authorizing a war that the United States had already decided on. The Iraq War is therefore a case in which the United States intended on ending the crisis through a brute force strategy but also planned on coercion failure to justify its actions. Though the administration sought to go to war with the Security Council's blessing, it was willing to endure the additional diplomatic and political costs and implemented its preferred strategy without its authorization.

As in the Gulf War, the asymmetric coercion model and coercion range provide a framework for understanding the dynamics of the 2002 to 2003 crisis. The absence of a threat to Iraqi state and regime survival explains why Saddam abandoned his policy of ambiguity over WMD, whereas the presence of such a threat to survival explains why he would not concede to regime change and the occupation of Iraq. The model does not, however, explain why the United States would choose a coercive strategy it wanted to fail. The diplomatic and political costs of abrogating the norm of first negotiating settlements through international institutions does explain why the Bush administration went to the UN before it went to war. The United States counted on coercion to fail and, in so doing, provide the justification, and thereby lower the costs for, the brute force strategy it intended. Unlike the Gulf War, for which this strategy worked quite well, in the lead-up to the 2003 invasion it backfired. This time Saddam conceded and the younger Bush administration found itself forced to absorb the diplomatic and political censure for an unjustifiable invasion and occupation.

The next chapter introduces two more cases of the United States in asymmetric interstate crises, this time pitted against Serbia and its president Slobodan Milosevic, over conflicts in Bosnia and Kosovo. Unlike the Iraqi cases, however, the United States preferred coercion over brute force war, and in both cases, though it took longer than anticipated, coercion eventually succeeded in bringing the crises to an end.

5 THE UNITED STATES VERSUS SERBIA
Bosnia and Kosovo

THE 1990S WITNESSED THE UNITED STATES involved in multiple long-term conflicts against much weaker nations. The previous chapter examined two crises in which Saddam Hussein's reign eventually ended with the U.S. invasion of Iraq. Against the Bosnian Serbs and Serbia, however, the United States chose coercion to achieve its foreign policy objectives in Bosnia-Herzegovina (hereafter referred to as Bosnia) and Kosovo, respectively.

In the Bosnian conflict, the Americans and their NATO allies challenged a weak Serbia that had just emerged from the dissolution of Yugoslavia. In addition, the United States faced ethnic Serbs who had formed a quasi-state in Bosnia. Though not internationally recognized, the Republika Srpska acted as a state, attempting to maintain control over its policy making and sovereignty over its territory.

The Bosnian civil war began with large-scale violence in April 1992, generating two million refugees. This energized the European Community and United Nations to form the International Conference on Former Yugoslavia (ICFY) in an ill-fated effort to negotiate a peace agreement. Not until February 6, 1994, when a mortar round killed sixty-eight civilians in Sarajevo's Markala marketplace, did the Americans step in. The United States took over negotiations and demanded the Serbs and the Muslim-Croat Federation accept a 49/51 partition of Bosnia, respectively. The United States also threatened military force, joining with NATO in enforcing a no-fly zone and supporting the arming, organization, and training of Croatian and Muslim armies. Economic sanctions, however, and not the threat of force convinced Serbia's

President Slobodan Milosevic to pressure the Bosnian Serbs to accept a peace deal. Even then, not until late summer of 1995 when the military advantage had turned in favor of Croat Muslim forces supported by NATO air power did the Bosnian Serbs finally concede. Denied Serbian military supplies and finding their western territories under threat, President Radovan Karadzic and General Ratko Mladic granted negotiating power to Milosevic. Milosevic then turned and struck a deal at the Dayton Peace Accords in November 1995, bringing the civil war to an end.

The asymmetric coercion model developed in Chapter 2 provides a method for analyzing the dynamics of U.S. and Serbian decision making. The model incorporates the sequential logic of asymmetric interstate conflicts, whereby the United States, with its great power advantage, has the luxury to choose whether to accommodate the weaker state. In the Bosnian conflict, the United States supported the implementation of sanctions and ICFY diplomatic efforts but refused to threaten or use its military forces until President Bill Clinton decided to act following the international reports of civilian casualties from the shelling of Sarajevo's Markala marketplace. Clinton escalated the conflict into a crisis for the United States by adopting a coercive strategy. Without vital national security interests at risk he assessed as too costly an alternative brute force strategy involving U.S. ground combat forces.

Nonetheless, the Bosnian Serbs rejected U.S. coercive demands as too onerous. The Clinton administration demanded the Bosnian Serbs hand over 30 percent of territory they had just gained through fighting and stipulated political concessions that would have dismantled the Republika Srpska. Threats to back up these demands were not yet credible, leaving the military balance of power in favor of the Bosnian Serb Army. Not until the United States lowered its demands for political reform and the Croat-Muslim offensive overran the contested territory did Karadzic and Mladic concede negotiating power to Milosevic. A coercion range finally materialized, wherein the Clinton administration preferred a peace deal, which avoided deploying U.S. troops into Bosnia to extricate NATO peacekeepers. The Bosnian Serbs also preferred conceding to demands that no longer threatened their survival over continuing a war that risked the demise of the Republika Srpska.

The Dayton Peace Accords ended the crisis in Bosnia but left unresolved political instability and ethnic unrest in the province of Kosovo. In the summer of 1998, fighting between Serbian forces and the Kosovo Liberation Army

(KLA) forced hundreds of thousands of ethnic Albanians to flee their homes. The Clinton administration acted quickly, insisting that Serbia withdraw the majority of its forces from Kosovo, a demand backed by the threat of NATO air strikes. Coercive diplomacy initially succeeded as Milosevic agreed in October 1998 to withdraw troops and to allow in the international monitors from the Organization for Security and Co-operation in Europe (OSCE). When the KLA began to occupy the positions being vacated by Serbian forces, Milosevic realized his mistake: He had, in effect, agreed to cede control over Kosovo. Refusing to relinquish Serbia's historic birthplace without a fight, Milosevic reversed his decision and ordered his troops back to their posts.

Shortly thereafter, in January 1999, the chief OSCE monitor accused Serb forces of a massacre in the Kosovo village of Račak. The Clinton administration responded with strong-arm talks at Rambouillet, France. Beyond the removal of Serbian forces from Kosovo, U.S. negotiators also demanded Serbia allow in NATO peacekeepers and, in three years' time, hold an international conference on the future of Kosovo. Coercive diplomacy failed, however, as Serbia refused, and in response NATO commenced Operation Allied Force (OAF) on March 24. These limited air strikes, initially planned for three nights, were insufficient to convince Serbia to concede its homeland territory. It would eventually take seventy-eight days and nights of air strikes with a Serbian economy on the brink of collapse before Milosevic acquiesced. Before he would concede, the United States reduced its demands, removing any reference to Kosovo independence and allowing Russian troops into Kosovo alongside NATO peacekeepers under a UN mandate.

Again, the logic of the asymmetric coercion model explains the dynamics of U.S. and Serbian decision making. The United States escalated the conflict into a crisis by choosing a coercive strategy with what appeared at the time to be moderate demands for Serbia to change its policies regarding its ethnic Albanian population. On concession, however, the subsequent seizure of territory by the KLA rendered the demands a claim on homeland territory. Likewise, threats of restricted air strikes against a small number of primarily military targets were eventually expanded to threatening Serbia's economy. Though coercive diplomacy failed on the commencement of NATO air attacks, a coercion range eventually emerged, wherein the United States preferred a coercive outcome to that of a brute force war, and Serbia preferred conceding Kosovo to further resistance.

Kosovo is an exceptional case in which a state conceded homeland territory while it still had the means to resist. Serbia's strategy of presenting the United States with a fait accompli by way of an ethnic cleansing campaign against the Kosovar Albanians backfired. Instead of bringing an end to the crisis, the continued attacks only strengthened U.S. resolve and NATO solidarity. With his strategy denied and Serbia's economy stressed by war, Milosevic conceded Kosovo to avoid further economic losses. Although British Prime Minister Tony Blair pressed Clinton to threaten a ground invasion, Milosevic conceded before the United States had taken steps to make such a threat credible.

Bosnia and Kosovo are both crises for which coercive diplomacy failed and the United States placed its reputation and prestige on the line for non-vital national security interests. The coercive strategies also achieved relatively meager foreign policy objectives and took longer and required a greater level of military intervention than anticipated. Despite this, coercion did succeed in obtaining the core foreign policy objectives set out by the Clinton administration. Analysis of these two asymmetric interstate crises is therefore useful for better understanding the limits to as well as what can be accomplished by coercion.

THE COLLAPSE OF YUGOSLAVIA AND THE ORIGINS
OF THE BOSNIAN CIVIL WAR

The origins of civil unrest in the Balkans can be found in the collapse of Yugoslavia (see Map 5.1 of the Western Balkans). Marshal Tito died in 1980 without having established a successor, an omission that led to growing tensions between the Yugoslavian republics throughout the 1980s. Slovenians and Croatians demanded a reduction in the influence of the federal government while the Serbians sought greater control over the federal government. Additionally, the province of Kosovo demanded that its status be elevated to that of a republic.[1]

By 1986, Serbia's future president, Slobodan Milosevic, a rising star in the Serbian Communist Party, had reached the position of party chief.[2] Ethnic tensions between Serbs and Albanians in Kosovo mounted in April 1987 when Milosevic appeared before a crowd of fifteen thousand denouncing the Kosovar Albanian demand for independence, "Yugoslavia does not exist without Kosovo."[3] Milosevic returned to Belgrade a national hero and leveraged his populist support to become president. He consolidated his power base

Map 5.1. Western Balkans.

by replacing leaders of the province of Voyjodina with men loyal to him and by forcing a new constitution on Kosovo.[4] In May 1989, he sought to control all of Yugoslavia through an additional alliance with Montenegro. Consequently, the relationship among Slovenia, Croatia, and Serbia grew increasingly tenuous.

In the fall of 1989, the collapse of communism in Eastern Europe brought the League of Communists in Yugoslavia down with it.[5] Milosevic responded by crafting a new Serbian constitution and repackaged the Communist Party as the Socialist Party of Serbia (SPS).[6] In December 1990, Slovenia declared its independence, and Belgrade responded by taking control of the Yugoslav People's Army (Jugoslovenska Narodna Armija, or JNA). In March 1991, Milosevic met secretly with Croatian President Franco Tudjman when, according to Tudjman, he and Milosevic agreed to the demarcation of borders between Croatia and Serbia and to the partitioning of Bosnia.[7]

On June 25, 1991, the republics of Slovenia and Croatia both declared independence in response to rising nationalism, the dismal economic conditions in Yugoslavia, and the growing assertiveness of Milosevic.[8] Slovenia's secession was relatively bloodless, following only ten days of fighting against an inept, Serbian-dominated JNA. Croatia was much less fortunate. The agreement between Tudjman and Milosevic did not prevent violence between Croatian forces and the JNA, which fought alongside ethnic Serbs from the eastern Croatian regions of Krajina and Slavonia.

The UN Security Council responded to this violence by adopting an arms embargo for all of Yugoslavia.[9] In November 1991, the Serbs and the Croats agreed to a UN-brokered truce and the deployment of nearly fifteen thousand UN Protection Forces (UNPROFOR) as peacekeepers.[10] Although the ceasefire provided a much-needed respite for Croatia to reorganize and rearm, it locked in territorial gains for the Serbs in Krajina and Slavonia. It also freed the Serbian-controlled JNA, now renamed the Army of Yugoslavia (VJ), to join Serbs in neighboring Bosnia in preparation for civil war.[11]

While Croats fought Serbs in Croatia's war, in Bosnia three ethnic factions vied for power. Parliamentary elections in November 1990 had split Bosnia along clear ethnic lines. Muslims constituted 44 percent of the population and were represented by the Party of Democratic Action (Stranka Demokratske Akcije, SDA) led by its founder and later Bosnian President, Alija Izetbegovic. At 31 percent, the Serbs elected the nationalist Radovan Karadzic as the head of the Serbian Democratic Party (Srpska Demokratska Stranka, SDS). As the most powerful party within Bosnia, the SDS enjoyed the support of both Serbia and the VJ. The SDS established its headquarters at Pale, a town just east of Sarajevo. The Croatian Democratic Union (Hrvatska Demokratska Zajednica, HDZ) represented the Croats, who made up 17 percent of the Bosnian population. Though the Bosnian Croats suffered under weak leadership, they received political and military support from Croatia's president, Franco Tudjman.[12]

After secession of Slovenia and Croatia, a tear appeared in the political fabric of Bosnia. Muslims and Croats favored independence over membership in a Serbian-dominated Federal Republic of Yugoslavia (FRY). Bosnian Serbs, however, preferred that Bosnia remain a republic within the FRY and refused to be a minority ethnic group in an independent Bosnia. On October 15,

1991, the Bosnian parliament, minus its Serbian contingent, voted for sovereignty but refrained from declaring independence.[13] In response, the Serbs declared their own Republic of Bosnia Herzegovina, later renamed the Republika Srpska, and on January 9, 1992, declared their independence from Bosnia.[14] Fearing civil war, the European Community (EC) initiated negotiations with the three Bosnian factions. The talks, however, were overcome by events as Bosnian Muslims and Croats voted for independence in a March 1, 1992, referendum boycotted by the Serbs.[15] From there, violence quickly spiraled into full-scale civil war.[16]

BOSNIAN CIVIL WAR 1992–1995

The three-year civil war in Bosnia left tens of thousands dead, displaced two million more, and introduced the term *ethnic cleansing* to the humanitarian intervention lexicon. The international community set the objectives of putting a stop to the killing while ensuring the existence of a multiethnic Bosnian state. They achieved this only through the leadership of the United States and a coercive strategy that combined economic sanctions with military force, the de facto partitioning of Bosnia, and the long-term commitment of U.S. ground troops.

By any measure, Bosnia is a complex and challenging case for analyzing asymmetric coercion. The conflict consisted of numerous actors from the Yugoslav republics and their ethnic-based political parties, all seeking independence. It also involved intervention from international institutions along with the great powers of the United States, Great Britain, France, and Russia. Within Bosnia, Muslims, Serbs, and Croats struggled for territory and sovereignty while the bordering states of Croatia and Serbia intervened both politically and militarily. International institutions stepped in early on in the conflict. The European Community and the United Nations formed the International Conference on Former Yugoslavia (ICFY) in an ill-fated attempt to bring about a peace agreement. As the conflict wore on, NATO countries became involved, sending in peacekeepers and conducting limited air operations to deter the Bosnian Serbs from further aggression. When ICFY efforts failed, the United States and Russia joined Great Britain, Germany, and France to form the Contact Group and began placing greater diplomatic and military pressure on the warring parties. Ultimately, the United States brought about a permanent settlement through a coercive strategy of sanctions, air strikes, and military support for the Croat Muslim offensive.

Though these efforts eventually achieved the United States' core ex ante objectives of maintaining a Bosnian state and ending the violence, Bosnia is hardly a shining success story for U.S. foreign policy. It took three violent years of civil war to bring the actors to the table and, to reach a settlement, the Clinton administration conceded to the partitioning of Bosnia and allowed the Bosnian Serbs, Croats, and Muslims to keep territory secured through ethnic cleansing. The United States was also complicit in circumventing a UN Security Council arms embargo in its efforts to arm the Bosnian Muslims. By the conclusion of the war, the Clinton administration had placed its reputation and U.S. prestige, along with that of NATO, at risk over nonvital national security interests. The United States then continued to the pay for its "success" by deploying peacekeeping troops to the region for nearly a decade.

The Bosnian Civil War can be divided into three coercive stages. In the first stage, from April 1992 to February 1994, the United States supported the ICFY peace effort.[17] This diplomatic initiative implemented economic sanctions and an arms embargo and deployed thousands of non-U.S. peacekeepers to the region but did not directly threaten military force. In the second stage, from February 1994 to April 1995, the United States took over negotiations by forming the Contact Group. During this period, the Clinton administration brokered a Muslim Croat peace agreement, resulting in the Bosnian Federation. The military balance of power in the region shifted in favor of the Bosnian Federation with U.S. support for its military buildup, accompanied by limited NATO air strikes against the Bosnian Serbs. The period then culminated in a temporary four-month cease-fire over the winter of 1994 until the spring of 1995.

In the third and final stage of the conflict, from May to November 1995, violence again erupted as Croatian forces assumed the offensive against the Serbs, gaining momentum in Croatia and then joining with Muslim forces in Bosnia in the summer of 1995. Alongside this offensive, the United States adopted a stick-and-carrot coercive strategy, threatening air strikes while offering a peace deal that would partition Bosnia by recognizing the Republika Srpska and would lift UN sanctions on Serbia. Coercion eventually succeeded, as a cease-fire was declared in mid-October and a political settlement reached at the Dayton Peace Accords in November 1995.

The First Coercive Stage: Vance-Owen Peace Plan

In April 1992, the civil war within Bosnia expanded into a regional conflict as both Croatia and Serbia moved forces into Bosnia. Following the Serb shelling

of Sarajevo, the United States and the European Community formally recognized Bosnia's independence in an attempt to constrain the fighting.[18]
However, the late May arrival of General Mladic, appointed by Milosevic as
commander of the newly formed Bosnian Serb Army, marked an escalation in
the violence. The shelling of Sarajevo intensified, prompting the UN Security
Council to impose economic sanctions against the Federal Republic of Yugoslavia (FRY), which by then encompassed Serbia and Montenegro, along with
the autonomous regions of Vojvodina and Kosovo.[19] The sanctions did little to
stop the fighting, and the superior-armed Serbs soon gained 70 percent of Bosnian territory through a campaign of ethnic cleansing, purging eastern and
northern Bosnia of nearly all of its Muslim and Croat population.[20] This, in
turn, generated a crisis for Europe as over two million refugees, roughly half
the Bosnian population, either fled the country or were internally displaced.[21]
Despite the enormity of the situation, without its vital national security interests at stake, the United States readily deferred to the Europeans and United
Nations to broker a peace agreement.

Under pressure to take action, the United Nations deployed 1,700
UNPROFOR (UN Protection Force) troops over the course of the summer of
1992 and, in the fall, authorized another 6,000 to secure the Sarajevo airport
and assist in the delivery of humanitarian assistance.[22] As the EC's diplomatic
efforts failed to produce a viable cease-fire, the United Nations joined forces
to form the International Conference on Former Yugoslavia (ICFY) in late
August 1992. The UN representative, former U.S. Secretary of State Cyrus
Vance, and the EC envoy, former British Foreign Secretary Lord David Owen,
cochaired the ICFY.[23] The comprehensive Vance-Owen peace plan revealed in
January 1993 proposed a decentralized Bosnian government divided into ten
provinces. Muslims, Croats, and Serbs would each retain majorities in three
of the provinces, while Sarajevo would become a separate open and demilitarized province.[24]

The Croats approved the Vance-Owen plan as it largely met their territorial ambitions by placing all three of their designated provinces along the
Croatian border. By contrast, the Bosnian (Muslim) government had serious
issues, both with the proposed constitution and with the division of territory.
President Izetbegovic held out hope for a military intervention by the United
States under the new Clinton administration. Albeit reluctantly, he did agree
to Vance-Owen to garner international support. He also correctly calculated
the Serbs would never agree to such a plan.[25]

The plan, in fact, produced a split between Serbs in Belgrade and those in Pale. Milosevic's vision for a Greater Serbia did not include Bosnia. For him, Vance-Owen required only that he change his policies, that is, stop support of the Bosnian Serbs and to pressure them to accept the peace plan. Although Milosevic did support the Bosnian Serbs in their early victories, by 1993 the deteriorating health of Serbia's economy and the negative impact of the UN-imposed sanctions concerned him more than their struggle. As a result of the dissolution of Yugoslavia and the ongoing wars in Croatia and now Bosnia, Serbia suffered from hyperinflation and a nearly 50 percent loss in GDP, conditions leaving it particularly vulnerable to sanctions.[26] With his powerbase anchored to his popularity and the support of the political elite in government and business, Milosevic prioritized Serbia's economy over the interests of Serbian nationalists who supported the Bosnian Serbs. Despite the unlikelihood that the Bosnian Serbs would concede, Milosevic began, at least publicly, to pressure them to accept Vance-Owen.[27] This move, condemned by Serb nationalists, led to the dissolution of the political alliance between Milosevic's Socialist Party of Serbia (SPS) and the Serbian Radical Party (SRS). In subsequent elections, he retained power by forming a more moderate government without the SRS.[28]

Despite his political interests, Milosevic did not press Karadzic for an agreement to the Vance-Owen peace plan until territorial gains in eastern Bosnia had been resolved.[29] Not until April 1993, when the Security Council voted to implement tougher sanctions against Serbia, did Milosevic finally begin to urge the Bosnian Serbs to sign.[30] At the time, however, they could afford to resist Milosevic as he had yet to secure effective political and economic pressure against them.[31] Seeing the provisions of Vance-Owen as a threat to the very survival of their newly formed republic, the Bosnian Serbs ignored Milosevic's demands and overwhelmingly rejected the plan in a mid-May 1993 referendum.[32]

As a coercive strategy, the Vance-Owen plan proved to be flawed as it mismatched the demands and threats made of the Bosnian Serbs. It called for the Serbs to relinquish 30 percent of their territory, much of which had just been gained in the past year's fighting. The provinces designated for the Serbs had been intentionally drawn to be noncontiguous to deny the viability of an independent Bosnian Serb state. The plan also expected the Bosnian Serbs to surrender their arms and accept a Bosnian constitution, concessions that would put an end to the Republika Srpska.

To back up Vance-Owen's demands the European Community and United Nations implemented economic sanctions but took no actions to signal a credible threat of military force. No major power committed its military to implement the plan. The lightly armed international troops deployed to Bosnia operated only under a UN peacekeeping mandate. More important, the balance of power lay in favor of the Serbs as they gained and held ground while the Croats and the Muslims expended their efforts fighting each other for what remained of Bosnia.

Still, coercive diplomacy nearly worked. In Athens on May 2, 1993, Karadzic caved under international pressure and signed on to the Vance-Owen peace agreement, though final approval rested with the Bosnian Serb National Assembly. Two weeks later, however, Mladic appeared before the Assembly, offering up an impassioned speech against Vance-Owen that prompted the assembly to reject the plan.[33] Coercive diplomacy ultimately failed because the demands placed on the Bosnian Serbs threatened the survival of the Republika Srpska. It would not be until the tides of war turned in the summer of 1995 and they found themselves on the defensive that Milosevic could leverage his political and media machine to exert sufficient influence over the Bosnian Serbs.

The Second Coercive Stage: The Contact Group
Though the United States desired a peaceful political transition for the former Yugoslavia, it viewed the situation as a primarily European issue without vital U.S. security interests at stake. As such, the United States supported the EC's diplomatic efforts, but, even as the humanitarian crisis unfolded over the summer of 1992, it would not commit military aid in the absence of a peace agreement.[34] The Bush administration maintained this policy even as presidential candidate Bill Clinton called for the United States to consider military force and a lifting of the arms embargo to help stop the ethnic cleansing.[35] In the fall of 1992 and throughout 1993, however, as thousands of European troops deployed to Bosnia as peacekeepers, the United States found its security interests indirectly connected through its commitments to its NATO allies.

The Vance-Owen proposal had died by September 1993. The ICFY changed tactics in the subsequent Owen-Stoltenberg and European Action plans and attempted to induce the Bosnian Serbs by redrawing the maps and conceding to the de facto partitioning of Bosnia.[36] Like Vance-Owen, however, these efforts demanded the more powerful Serbs relinquish territory without the credible threat to compel them to do so.

The dynamics of the conflict changed on February 6, 1994, however, when a single mortar round landed in Sarajevo's Markala marketplace, killing sixty-eight Muslim civilians.[37] The attack and subsequent media coverage caused the Clinton administration to escalate the conflict into a crisis by making two changes to its foreign policy. Militarily, the United States would now threaten NATO air strikes should the Serbs not cease the shelling and remove their heavy weapons from around Sarajevo.[38] The United States also demonstrated its heightened resolve by its willingness to enforce the UN's previously man-dated no-fly ban.[39] On the political front, the Clinton administration com-menced negotiations to end the fighting between the Croats and Muslims, reaching a cease-fire on February 22 and then forming the Bosnian Federation at the end of March.[40] This proved a key diplomatic achievement that would eventually shift the balance of power in the Federation's favor as both Croats and Muslims rearmed and refocused their attention against the Serbs.

In late March and April 1994, Serbs attacked the Bosnian Muslim enclave of Gorazde, one of six designated safe areas. In an effort to deter further at-tack, the United Nations approved three limited NATO air strikes.[41] The Serbs retaliated by shelling the town, downing a NATO fighter, and taking 120 UN personnel hostage.[42]

The Gorazde crisis brought an end to ICFY efforts, and on April 24, 1994, the newly formed Contact Group, led by the United States and to a lesser ex-tent Russia, joined by representatives from France, Germany, and Great Brit-ain replaced the ICFY.[43] The Contact Group differed from previous efforts by employing bilateral negotiations with the Americans now charged with bring-ing the Croats and Muslims to the table and the Russians responsible for the Serbs. The Contact Group announced its peace plan in mid-May, proposing a federated Bosnia, partitioning it with 51 percent of the territory going to the Muslim Croat Federation and 49 percent to the Bosnian Serbs.[44] Unlike Vance-Owen, this new proposal significantly reduced the level of demands on the Bosnian Serbs by allowing them to retain their government, to keep their heavy weapons, and to form independent diplomatic relations with Serbia.

The United States advocated a "lift and strike" strategy, threatening to lift the arms embargo on the Bosnian (Muslim) Army and strike the Bosnian Serb Army with NATO air power to compel the Bosnian Serbs to accept the terms of the Contact Group's plan. The United States also wished to increase economic sanctions on Serbia to further motivate Milosevic to pressure Pale. The Russians, for their part, approved of the partitioning plan but resisted the other measures. In the end, the plan, like previous efforts, lacked sufficient coercive leverage.[45]

Predictably, the Bosnian Serb parliament rejected the Contact Group plan in July 1994.[46] Though Bosnian (Muslim) and Croatian forces had taken the initiative, they had yet to make serious gains against the more powerful Serbs, who still held 70 percent of Bosnia and had little reason to give up nearly a third of their territory. Even under pressure from their allies, Russia and Serbia, they continued to resist.

On August 4, 1994, after the Bosnian Serb parliament rejected the Contact Group's proposal for a third time, Milosevic implemented economic and diplomatic sanctions against the Bosnian Serbs.[47] Milosevic continued to publicly pressure Karadzic and Mladic, even going so far as to order the blockade of roads, to cut economic ties, and to freeze the pay of Bosnian Serb military officers.[48] These actions had a decided effect on both men. Mladic traveled to Belgrade in an effort to make amends with Milosevic. For his part, Karadzic initiated contact with former U.S. President Jimmy Carter, in an effort to influence the United States, a move that would later result in a temporary cease-fire.[49]

Milosevic could not, however, bring the Bosnian Serbs to concede to the Contact Group's proposal and give up their hard-fought gains in homeland territory. They failed to fully appreciate, however, what effect a withdrawal of Serbian support would have on the balance of power in the region, particularly for their fellow Serbians in Croatia. The fate of Krajina and Slavonia, however, had long been decided by Milosevic and Tudjman in their secret meeting back in March 1991.[50] The Krajina Serbs, who relied even more heavily on Serbia's backing to deter Croatian forces, could not defend against a Croat offensive. The Bosnian Serbs had not considered nor prepared for what would happen if Krajina fell. This would create a second front, threatening undefended western Bosnia including their largest city of Banja Luka. So long as they misperceived that the balance of power in Bosnia would remain in their favor, the Bosnian Serbs felt insufficient pressure to concede to a 51/49 partition while they still held 70 percent of Bosnia.

Unable to form an appropriate response in the face of the obstinate Bosnian Serbs, the Contact Group did little. The United States considered a unilateral lifting of the arms embargo for the Bosnian Federation Army but faced the opposition of Great Britain and France, who feared such action would increase the risk to their peacekeeping forces. The Clinton administration instead adopted an opaque policy. Although the United States did not openly arm the Croats or Muslims, it did provide training and encouraged clandestine arms shipments from Muslim countries.[51]

The air power portion of the U.S. strategy did not materialize until November 1994. During the previous summer, the Bosnian army had advanced into northwest Bosnia, forcing Serbs out of the UN-designated safe area of Bihac.[52] The Serbs mounted a counterattack that included air strikes and, by mid-November, was in position to overrun the town. In response, the UN Security Council authorized NATO air strikes against the Serb surface-to-air missile sites and the Serb airbase at Udbina in neighboring Krajina.[53] The results of the strikes were dismal. The Serbs retaliated in a similar fashion as to previous air strikes by detaining over 200 UN personnel near Sarajevo. Fearing the Serbs would keep their personnel hostage, NATO backed down.[54]

The Serbs, however, continued to detain and harass UN troops; by December 1994, France, the largest contributor of UN peacekeepers, called for NATO to begin planning for a UNPROFOR withdrawal.[55] This led to the NATO Council endorsement of OPLAN 40-104. This plan called for the deployment of up to 20,000 U.S. ground troops to assist in a withdrawal from Bosnia. The United States agreed to send these troops to the region, should such action become necessary.[56] At the same time, Karadzic, under increasing pressure to seek a peace agreement, initiated a cease-fire proposal through Jimmy Carter. Karadzic agreed to reopen the Sarajevo airport, to allow the movement of humanitarian aid, and to stop the harassment of UN personnel in return for a four-month cease-fire and the recommencement of serious negotiations.[57] Though it largely held until May 1995, this Serbian-initiated cease-fire did not produce a permanent peace and had the unintended effect of providing an opportunity for the Croatian and Bosnian Federation Armies to prepare for a summer offensive.

Though U.S. efforts at coercion failed at this juncture, the seeds for ultimate success were sown in the formation of the Croat-Muslim Bosnian Federation. This alliance, supported by NATO air power, would eventually change the balance of military power on the ground in Bosnia, proving a viable threat to Bosnian Serbs and finally bringing both Karadzic and Mladic under the control of Milosevic.

The Third Coercive Stage: The Croatian Offensive and the Dayton Accords

On May 1, 1995, after three years of relative calm, war returned to Croatia. The Croatian Army pushed into western Slavonia, meeting light resistance

and, within a week, had retaken the region, though Serbs still controlled eastern Slavonia and Krajina.[58] The strong showing by the Croatian Army was
attributable to two factors. First, the Croatian Army proved a much better
fighting force than what it had been in 1991. The Clinton administration had
supported Croatia's military buildup by encouraging arms embargo violations
and allowing retired senior U.S. officers to advise the Croatian Army.[59] Second, and more important, Milosevic had withdrawn Serbian support from the
Krajina Serbs.[60]

The cease-fire likewise collapsed within Bosnia as fighting broke out
around several cities, including Sarajevo. Following three weeks of hard fighting, UNPROFOR issued an ultimatum on May 25, 1995, calling for both the
Bosnian Serb and Bosnian (Muslim) Armies to refrain from employing their
heavy weapons near Sarajevo or face NATO air strikes.[61] When the Serbs refused to comply, NATO struck an ammunition depot near Pale. This elicited
a now predictable response from the Bosnian Serbs, who shelled five of the six
safe areas and seized 400 UN personnel, this time displaying them handcuffed
as human shields in front of potential NATO targets.[62] By June 10, the Bosnian
Serbs had clearly won the standoff. To secure the release of the remaining
hostages, the UN announced it would "return to the status quo" and "abide
strictly by peacekeeping principles until further notice."[63]

In June 1995, France and Britain began sending mixed signals as to their
resolve over Bosnia. While French and British diplomats publicly questioned
how long they would continue to support UNPROFOR, their militaries deployed heavily armed units to the region as a new rapid-response force. It remained unclear, however, how this additional ground power would be used,
that is, whether to provide additional military might to back up UN ultimatums or to facilitate a withdrawal.[64] At the same time, the hostage crisis and
a disagreement over terms for suspending the UN sanctions scuttled bilateral
talks between the United States and Serbia. Milosevic, bolstered by the recent Bosnian Serb victory, demanded the sanctions be permanently lifted and
balked at U.S. insistence that the Security Council retain the right to reimpose
them.[65]

In July 1995, in response to raids by Muslim forces staging out of the
UN safe area of Srebrenica, Mladic ordered the shelling of the city. With the
United Nations unwilling to authorize NATO air strikes, the lightly armed
and outnumbered UNPROFOR troops could do no more than withdraw as
the Bosnian Serb forces overran the city and conducted the mass killing of

over 7,000 Muslim men.[66] This gruesome attack, along with the fall of Zepa two weeks later, shocked the United States and Western Europe into action.

Clinton acknowledged to his national security team that the current U.S. foreign policy had proven untenable and that the situation in Bosnia made the United States look weak and did "enormous damage to the United States' . . . standing in the world."[67] This newfound resolve was, in part, due to his belated recognition that, regardless of the outcome in Bosnia, he had already committed 20,000 ground troops to deploy to the region, whether to enforce a peace agreement or to assist in a potentially violent UNPROFOR withdrawal.[68] Clinton therefore pressed for the peace agreement as, politically, he could ill afford to deploy troops to enforce a failed foreign policy in the midst of his 1996 reelection bid.[69]

A July 21, 1995, conference, hastily convened in London and attended by NATO leaders, a Russian representative, and the UN General Secretary's envoy, produced two fundamental changes to NATO policy. First, they drew a line in the sand declaring that an attack on Gorazde, the last UN safe area in eastern Bosnia, would "be met by substantial and decisive air power."[70] They later expanded this protection to include the other remaining safe areas of Sarajevo, Bihac, and Tuzla.[71] Second, NATO air strikes had been hampered by a "dual key" approval process, which required authorization by UN civilian officials and NATO military officials. UN Secretary General Boutros-Ghali now agreed to delegate UN strike authority out of civilian hands and into those of the overall military commander for UNPROFOR, French Lieutenant General Bernard Janvier. This significantly streamlined the NATO–UN air strike approval process.[72]

The Clinton administration then announced a new "End Game" strategy in early August 1995.[73] The plan called for a diplomatic initiative to reinvigorate the Contact Group's proposal, this time adding the threat of air strikes against the Bosnian Serbs if they rejected the plan.[74] Unlike a similar U.S. proposal in 1993, the Europeans now agreed to the expanded role of NATO air power over the tepid objections raised by Russia.[75]

A shift in the balance of military power within the Balkans in favor of Croatia, the Muslim–Croat Bosnian Federation, and NATO, all aligned against the Serbs accompanied the stiffening of U.S. resolve. On August 4, 1995, the Croatian Army commenced a new offensive, ethnically cleansing over 200,000 Serbs from Krajina. This left eastern Slavonia the sole Croatian territory still under Serbian control.[76] The stunned Bosnian Serbs not only witnessed the collapse of Krajina, which opened a new front to their west, but also saw the influx of thousands of Serb refugees. As with western Slavonia, Milosevic withheld military support and instead blamed Krajina Serb leadership for failing to reach a settlement with Croatia's President Tudjman.[77] The Croat offensive had a rippling effect in northwest Bosnia, relaxing the Serb stranglehold on the Bihac pocket and allowing pinned-in Bosnian Muslim units to break out of the city. Meanwhile in central Bosnia, both Croatian and Muslim Federation troops began attacking Serb positions.[78]

As the tide of war turned against the Bosnian Serbs, their leadership began to show signs of strain, and their solidarity publicly unraveled. Karadzic blamed the recent losses in western Bosnia on Mladic, and he moved to relieve Mladic of command.[79] Having originally been appointed to his position by Belgrade, however, Mladic now traveled to Belgrade to confer with Milosevic. In the end, Mladic refused to step down and declared Karadzic's order illegal. Backed by the entire Bosnian Serb General Officer Corps, Mladic won the day and forced Karadzic to reverse his decision.[80]

Meanwhile, the United States' newly appointed chief negotiator in the Balkans, Richard Holbrooke, sought to bypass both Bosnian Serb leaders altogether. He met with Milosevic in Belgrade on August 17, 1995, and announced that the United States would no longer negotiate with the Bosnian Serbs, dealing instead with Milosevic alone.[81]

State and Regime Survival Explains Milosevic Concession

Milosevic conceded to the Contact Group's peace plan and agreed to negotiate on behalf of the Bosnian Serbs as such a concession did not threaten the survival of Serbia. Bosnia had never been part of Serbia and had maintained an autonomous status as a republic for over a century prior to the dissolution of Yugoslavia in 1991. The Contact Group's proposal did not infringe on Serbia's domestic sovereignty over of its homeland territory or on its international sovereignty over foreign policy decision making. A peace agreement would, however, eliminate the dire threat to Serbia's economy from the UN's sanctions, in place since 1992.

Conceding to international demands did not threaten Serbia's survival, nor did it rouse the threat of domestic opposition to the Milosevic regime. His Socialist Party of Serbia (SPS) firmly controlled the government, and no armed groups within Serbia had the capability of violently overthrowing the regime. Back in November 1990, Milosevic had sealed an alliance between the

SPS and the Yugoslavian Army. This successful subjugation of the military to civilian control not only reduced the chances of a military coup but also deterred the formation of armed opposition groups that could threaten revolt.[82] As a result, unlike Saddam, who faced the March 1991 uprising by Shia opposition groups, Milosevic did not foresee the threat of an insurgency. His SPS was vulnerable at the ballot box, however. Although the SPS had won elections since Milosevic's rise to power, it had done so by an ever-decreasing margin. Milosevic's mishandling of Serbia's economy had resulted in a growing displeasure over his leadership. Conceding to U.S. demands assisted Milosevic by bringing the costly Bosnian Civil War to a close and by lifting the debilitating UN sanctions.

As the leader of a single party regime forced to form coalitions to retain power, Milosevic would likely suffer larger audience costs for making concessions than Iraq's Saddam Hussein, the leader of a personalist regime.[83] Milosevic, however, undertook two actions to undercut the backlash generated by his pressuring the Bosnian Serbs to accept a peace agreement. First, back in February 1994, Milosevic had begun employing his propaganda machine to place the blame for Serbia's growing economic crisis on the unwillingness of the Bosnian Serbs to accept a peace deal. Second, he had changed his policy incrementally, by implementing, but weakly enforcing, economic sanctions on the Bosnian Serbs, then enforcing those sanctions more strictly over time and, finally, threatening to withdraw all Serbian military support for their war effort. These actions deflected audience costs away from Milosevic.

Karadzic and Mladic Concede Negotiating Power to Milosevic

Two Serbian artillery shells killed thirty-seven in another attack on Sarajevo's Markala marketplace on August 28, providing the pretext the United States had been waiting for to commence preplanned NATO air strikes.[84] Karadzic and Mladic were still stunned over the collapse of Krajina; with air strikes imminent, they felt particularly vulnerable. Summoning the bickering pair to Belgrade on August 30, Milosevic finally wrested from them the authority to negotiate on behalf of the Bosnian Serbs.[85]

The ceding of negotiating power to Milosevic proved to be a critical juncture in bringing about a peace agreement. Two related factors explain this reversal by Karadzic and Mladic, the buildup of Croat and Muslim military forces and Serbia's withdrawal of military support. The balance of military power between the Bosnian Serbs and the Federation began to shift with the collapse of Krajina, which left the Bosnian Serbs' western flank exposed to the Croat Muslim offensive. Mladic, however, could not reinforce Banja Luka, as his heavy weapons were dedicated to the siege of Sarajevo. He feared Muslim troops would occupy the positions vacated by a withdrawal from Sarajevo, and NATO air power threatened the transport of weapons from eastern to western Bosnia.[86]

Though the Bosnian Serbs' situation in the west had become serious, it was not yet dire. Mladic therefore would not concede to NATO's military demands to remove his weapons from Sarajevo until he received assurances that Russian peacekeeping troops would replace his forces.[87] Such a deployment would prevent the Muslim forces in Sarajevo from further territorial gains and provide a buffer between the Bosnian Serb and Federation forces in central Bosnia. The Serbs in Croatia had employed this tactic in 1991, when UNPROFOR troops deployed to eastern Croatia, thus freeing the Serbs to reinforce Bosnia.[88]

Milosevic pulling the plug on Serbia's support of the Croatian and Bosnian Serbs proved to be the second factor in Karadzic and Mladic's concession of negotiating power. Serbia had already secured its border and reduced the flow of goods into Bosnia. More important, the fighting during the summer of 1995 had nearly exhausted the Bosnian Serb Army, leaving them dependent on Serbia to continue their military operations. Demoralized, Karadzic and Mladic had finally concluded that they were losing and needed Milosevic to avoid a defeat that could cost the Bosnian Serbs much of their territory and threaten the survival of the Republika Srpska.

Operation Deliberate Force, August 30–September 14, 1995

Following the August 28, 1995, shelling of the Markala marketplace, the Clinton administration immediately called for the United Nations to approve the implementation of a NATO air operation.[89] Following a one-day delay to secure the withdrawal of the remaining UNPROFOR troops from Gorazde, Operation Deliberate Force commenced on August 30. Over the next two days, NATO launched 372 strike sorties against the Bosnian Serb Integrated Air Defense System, artillery positions, ammunition depots, and command and control centers located near Sarajevo and Pale in southeast Bosnia.[90] In addition, the UN's Rapid Reaction Force fired over a thousand artillery rounds against Bosnian Serb positions near Sarajevo and in western Bosnia.[91]

At the onset of air strikes, General Janvier dispatched messages to Mladic with three conditions for halting the bombing: Cease military operations threatening the four safe areas, withdraw all heavy weapons from Sarajevo, and cease hostilities throughout Bosnia.[92] On August 31, Janvier requested from NATO a twenty-four-hour bombing pause to meet with Mladic but then rejected Mladic's conditional acceptance to withdraw under a guarantee that the Bosnian (Muslim) Army would not take over the vacated territory.[93] The bombing pause was further extended until September 5, when confusion arose over who had the authority to speak for the Serbs. Milosevic, who had yet to fully exert control over the Bosnian Serbs, made an attempt to accept the UN terms by contacting the secretary general's senior civilian envoy, Yasushi Akashi, while Karadzic sent a conciliatory message through Jimmy Carter.[94]

Regardless of diplomatic efforts made, by September 5 there was no evidence of Bosnian Serb heavy weapons being removed from Sarajevo.[95] NATO air strikes recommenced, this time accompanied by a revised ultimatum for the Bosnian Serbs to cease their attacks on Sarajevo and the other safe areas, to immediately withdraw their heavy weapons from the twenty-kilometer exclusion zone around Sarajevo, and to allow the free movement of UN and nongovernment organization personnel in and out of Sarajevo.

In addition to military activity, U.S. diplomacy had begun to make inroads. In Geneva on September 8 Croatian, Bosnian, and Federal Republic of Yugoslavia (Serbia and Montenegro) foreign ministers agreed to basic principles for a Bosnian settlement that recognized the international borders of Bosnia and formed two political entities within Bosnia, the Muslim Croat Federation of Bosnia and the Serbian Republika Srpska with territory divided 51 percent to 49 percent, respectively.[96]

On September 9, the Croatian and Bosnian Federation Armies launched a coordinated ground offensive into western Bosnia.[97] The same day, Mladic informed Janvier that the Bosnian Serb Army would now meet the UN ultimatum. The two met, along with Milosevic, in Belgrade the following day.[98] Meanwhile, NATO expanded its air operations into western Bosnia supporting the Croat Muslim ground offensive, which gained momentum and threatened Banja Luka.[99]

On September 13, Richard Holbrooke met with Milosevic, Karadzic, and Mladic in Belgrade, where the Serbs agreed to the basic principles signed in Geneva the previous week.[100] In addition, Mladic agreed to end the siege on Sarajevo and to remove all heavy weapons, this time receiving assurances that

Russian UNPROFOR troops would occupy the positions being vacated by the Serbs.[101] In return, NATO suspended air strikes the next day, on September 14, initially for seventy-two hours and then permanently, once the Bosnian Serbs were deemed in compliance.[102]

Permanent Cease-Fire

Although NATO suspended air strikes, the Croat Muslim ground offensive continued, forcing the Serbs to relinquish large portions of western Bosnia.[103] Holbrooke, who preferred to enter formal peace negotiations with the ground reality closely matching the Contact Group's 51/49 partitioning, encouraged the Croats to continue seizing territory but cautioned against taking Banja Luka.[104] By September 19, without the support of NATO air power, the offensive had begun to lose steam. The Bosnian Serbs in the easily defendable mountainous terrain hampered forces approaching Banja Luka from the south while Croatian forces from the north took combat losses as they attempted to cross into Bosnia over the Una river at Dubica.[105]

By October, the Croat Muslim offensive had stalled, and the Bosnian Serbs showed signs of mounting a counteroffensive.[106] On October 5, with the Federation and Bosnian Serbs each controlling roughly half of Bosnia, Holbrooke finally secured a cease-fire agreement among Tudjman, Izetbegovic, and Milosevic. Clinton then announced that the cease-fire would officially commence on October 10, with peace talks to take place later in the United States.[107]

Dayton Accords

Negotiations began on November 1, 1995, at Wright-Patterson Air Force Base in Dayton, Ohio, and concluded three weeks later with a formal signing of the Dayton Accords in Paris on December 15.[108] As with the October cease-fire agreement, Holbrooke conducted the talks primarily with Tudjman, Itzetbegovic, and Milosevic. Neither Karadzic nor Mladic, now internationally indicted war criminals, attended.

An early agreement between Milosevic and Tudjman resolved the remaining issue of the Croatian war, that is, the return of eastern Slavonia to Croatia. Milosevic agreed to turn over the region; in return, Tudjman supported the Bosnian peace process.[109] The more difficult and time-consuming aspect of the talks lay in defining the interentity border to separate the Federation from the Republika Srpska.

To gain an agreement, Milosevic conceded on two key territorial issues. First, he agreed to give up Serb-held sections of Sarajevo and territory in eastern Bosnia to provide the Federation with a secure access route to Gorazde. Second, he agreed to delay for a year a decision over the Brcko corridor and to ultimately submit the issue to international arbitration.[110] In return, the Republika Srpska retained 49 percent of Bosnian territory and received recognition as a separate political entity within Bosnia with the right to directly interact with Serbia. It retained its military, though its heavy weapons would be assigned to UN-monitored cantonment areas. As promised, Milosevic delivered the cooperation of the Bosnian Serbs when he traveled to Bosnia the following week and secured the signatures of both Karadzic and Mladic.[111] For his efforts, Milosevic finally succeeded in having UN sanctions against Serbia lifted.

Conclusion of the Bosnian Civil War:
Survival Concerns Explain Coercion Success

By the October 1995 cease-fire, conceding to the Contact Group plan no longer threatened state survival for the Republika Srpska. By contrast, the earlier Vance-Owen, Owen-Stoltenberg, and European Action plans had all threatened to take away territory and the Bosnian Serb government's control over its foreign and domestic policy making. The Vance-Owen plan proved particularly onerous as it, in effect, demanded regime change by not even recognizing the Bosnian Serb government. The subsequent Owen-Stoltenberg and European Action plans were less threatening for the Republika Srpska. Yet, although these proposals did allow for the de facto partitioning of Bosnia, they still required major territorial concessions.

When introduced in May 1994, the Contact Group's proposed 51/49 territorial split, likewise, required the Bosnian Serbs to concede land that they had fought for and held for three years. By the fall of 1995, however, the situation on the ground had changed, and the Croat and Muslim offensive had reduced Bosnian Serb-held territory to roughly half of Bosnia, now reflective of the plan's partitioning. Conceding to the Contact Group's plan no longer threatened the survival of the Republika Srpska, as this peace agreement provided international recognition of the republic, allowed the Bosnian Serb Army to maintain its heavy weapons at cantonment sites, and enabled special economic and diplomatic ties to be established with Serbia.

The Contact Group plan did not demand regime change, nor were domestic opposition groups capable of overthrowing Karadzic and his Serbian Democratic Party. The logic of omni-balancing applies to regimes threatened by violent domestic opposition. Such armed groups look for signs of weakness in a regime as a trigger for revolt. In this case, although the Bosnian Serb government confronted multiple fronts, that is, NATO, the Bosnian Federation, Serbia, and, at times, by internal dissension, there is no evidence that it was threatened by revolt.

The final analysis of survival requires an assessment of whether the regime leadership of Karadzic would have suffered significant audience costs for conceding to a peace agreement. In May 1993, Karadzic had been pressured internationally to end the war; by the fall of 1994, he was beginning to feel domestic pressure for a cease-fire. Turning to a third party, Karadzic solicited Jimmy Carter to negotiate a four-month cease-fire in an effort to buy time. Unfortunately for the Bosnian Serbs, Karadzic was unable to garner a permanent peace agreement that would allow the Republika Srpska to retain the territory it held. The cease-fire, instead, proved more beneficial to Croatian and Muslim forces, granting the two armies time to make preparations for a summer offensive.

Indeed, the Croatian offensive against Krajina in July 1995 and the withdrawal of Serbian support led to the collapse of the Krajinian Serbs, creating a ripple effect across northwest Bosnia with the influx of thousands of refugees. In central Bosnia, both Croatian and Muslim Federation troops simultaneously began attacks on Serb positions.[112] Further Federation advances in August, impending NATO air strikes, and Milosevic's threat to withdraw all military support to Bosnia had combined to place the survival of the Republika Srpska on the line. Concession at this point no longer generated as high an audience cost for Karadzic as before, when the Bosnian Serbs had had the military advantage. Even so, he attempted to deflect this cost by first blaming Mladic for the military defeats in western Bosnia and then Milosevic for withholding support.[113]

Conceding to a permanent cease-fire and peace agreement did not risk the survival of the Republika Srpska, of Karadzic, or of his regime. The reverse was actually the case, as the survival of the Republika Srpska would have been threatened had he not conceded. Survival concerns better explain the decision making of the Bosnian Serbs than do the conventional explanations for war introduced in Chapter 2. Misperception, miscalculation, and uncertainty do, however, explain why coercive diplomacy failed and why it took over three years for the United States to achieve a successful coercive outcome. The bungled diplomatic efforts of the ICFY in the Vance-Owen peace proposal resulted in high level demands for territory and regime change without a corresponding threat of military force. Miscalculation by the Bosnian Serbs that Milosevic would continue to support the Bosnian Serbs and their surprise at the Krajinian Serb's sudden collapse explain Karadzic and Mladic's subsequent surrender of negotiating power to Milosevic in late August 1995. Though these conventional explanations help us understand the delays in reaching a negotiated settlement, they do not explain why perceptions and calculations changed and why uncertainty was reduced to the point where a settlement could then be reached.

In addition to misperception, miscalculation, and uncertainty, the commitment explanation for war expects coercion to fail when the United States, as the powerful challenger, cannot convince the weak target that a concession will lead only to further demands. In this case, however, as in the previous chapter on Iraq, there is no evidence that Serbia or the Bosnian Serbs were concerned that the United States would not live up to its promises. Four factors influenced the Bosnian Serb calculation that commitment problems would not be an issue, that is, that the Americans would make no further demands. First, the Clinton administration included Russia in the Contact Group and charged it with bringing the Serbs to the negotiating table. The Bosnian Serbs placed more trust in the Russians, as witnessed by Mladic's refusal to remove his heavy weapons from Sarajevo until assured that it would be Russian troops that occupied their vacated positions. Participation by Russia also increased potential diplomatic costs for the United States, had it decided to renege on the agreement. The United States would have suffered strained relations with Russia, particularly because Russia had placed its reputation on the line.

Second, the involvement of Milosevic helped reduce commitment concerns for the Bosnian Serbs. His negotiating power for the Republika Srpska was contingent on his holding fast to the 51/49 split of Bosnian territory. At Dayton, Milosevic made it clear that he could not take an agreement back to the Bosnian Serbs that contained less. This point was non-negotiable and resulted in last-minute trades of worthless mountainous terrain in western Bosnia in exchange for broadening the Gorazde corridor in the east.[114] Though Milosevic also made concessions with regard to Sarajevo and Goradze and deferred the Brcko corridor to international arbitration, he came away with the promised 49 percent of Bosnian land.

Third, NATO suspended air strikes once Mladic agreed to remove the heavy weapons from Sarajevo.[115] So long as the Bosnian Serb Army fulfilled the terms of the cease-fire, it would be diplomatically costly for the United States to recommence the strikes. This loss of air support also weakened the Croat-Muslim offensive, thereby limiting the availability of military force to back up additional demands the United States might have otherwise contemplated.

Finally, the commitment problem explanation for coercion failure presumes that a target state reveals itself to be weakly resolved when it makes a concession. A target may, however, be able to mitigate this appearance of weakness by first enduring some punishment. Having withstood twelve days of NATO air strikes, the Bosnian Serbs appeared tougher and more resolved than they would have, had they conceded prior to the air strikes.

For the Bosnian Civil War, survival better explains the actual outcome of the crisis. The Bosnian Serbs conceded only at the point when the Contact Group's proposals no longer threatened their survival and further resistance would have risked survival, as they had just witnessed with the Krajina Serbs. Milosevic's concessions likewise did not threaten his survival, though the continuation of sanctions on Serbia took a toll on his political power and motivated him to sign the Dayton Accords. Dayton would bring an end to the Bosnian Civil War, but Milosevic's refusal to also include discussions over the future of the province of Kosovo would lead to a second crisis for the United States.

Post-Dayton Serbia: The Question of Kosovo

Milosevic emerged from Dayton with two achievements, the lifting of UN sanctions and an international reputation as a peacemaker. Back home, however, years of bloodshed over Croatia and Bosnia and economic decline had taken a toll on his popularity and, after the November 1996 elections, his power had been substantially weakened. Prior to the election, his political coalition had consisted of his Socialist Party of Serbia (SPS) and his wife Mirjana Markovic's smaller Yugoslavia Left (JUL) party. To retain a majority of seats in the Yugoslav Federal Assembly, he was forced to include in his government the former opposition party, the liberal New Democracy (ND). Even so, the remaining opposition, loosely banded under the Together (Zajedno) coalition, a group that had in the past been unable to pose a credible challenge, had now gained ground in local elections and taken control of Belgrade and several other municipalities.[116] A stunned Milosevic attempted to falsify the election results but finally relented in the face of large-scale public protest.

Milosevic remained in power but felt increasing political pressure from both the Together coalition on the left and nationalists on the right. In 1998 he moved to co-opt the nationalists by bringing into his coalition two of their leading voices: Vuk Draskovic, founder of the Serbian Renewal Movement

(SPO), and Vojislav Seselj, founder of the extremist Serbian Radical Party (SRS).[117] With this shift toward nationalism, Serbia would become less tolerant of ethnic Kosovar Albanians and their demands for independence.

Though Dayton ended the Bosnian Civil War, the peace conference avoided discussing the ethnic tensions that had long plagued Kosovo (see Map 5.2). Kosovar Albanians, led by Ibrahim Rugova and his Democratic

Map 5.2. Serbia.

Alliance of Kosovo (LDK), had for many years opposed Serbia through a strategy of peaceful resistance. Bosnia had proven, however, that only violence could goad the international community into action. Following Dayton, the Kosovo Liberation Army (KLA, also known as the UCK, Ushtria Clirimtare e Kosoves), a small guerilla movement formed in the early 1990s, began to gain support outside of the capital of Pristina.[118] In late 1996, the military strength of the KLA was also greatly enhanced by a bonanza of small arms made available by the collapse of the neighboring Albanian government. Cheap weapons flowed from Albanian armories into the eager hands of KLA fighters.[119]

By early 1998, the KLA emerged from the shadows, waging attacks on Serbian civilians, police, and Kosovar Albanians cooperating with the Serbs.[120] Serbian police responded with a series of heavy-handed raids on the homes of KLA leaders, including the highly publicized March 5 attack on the Jashari clan in Prekaz, which, according to conflicting reports, left forty to sixty dead, among them women and children.[121]

Following the attack, the KLA expanded rapidly. From an informal organization of approximately 200, the KLA's membership grew to a thousand in the course of two weeks. The KLA also reorganized, adopting a more conventional military structure, forming a headquarters, and establishing seven zone commanders responsible for the brigades and operations assigned to them.[122] Most of the new volunteers had no weapons or military experience, and many traveled to Albania to procure weapons and receive a minimal level of training. In addition to recruits, funds from the Kosovar diaspora began to flow quickly into the region, with close to $200,000 raised by the KLA in just days following the Prekaz attack.[123]

THE CRISIS OVER KOSOVO

The killings of the Jashari clan in Prekaz in March 1998 triggered a crisis for the United States. The Clinton administration's experience in Bosnia told them that ethnic violence in the Balkans could easily spiral out of control, and only tough, decisive U.S. diplomacy could stop it. In addition, key administration officials had also mistakenly taken from Dayton that Milosevic would back down once threatened with military force. The United States therefore rejected further accommodation of Milosevic's policies toward Kosovo and instead adopted a coercive strategy threatening NATO air strikes. As with Bosnia, in the absence of vital national interests President Clinton did not consider a brute force invasion.

In March, Secretary of State Madeleine Albright, blaming Milosevic for the surge in violence, urged the Contact Group to impose sanctions.[124] She then dispatched Richard Holbrooke in May to deliver a stern warning to Milosevic to end the fighting in Kosovo.[125] Neither the threat of sanctions nor diplomacy, however, had much impact on the ground, as the fighting continued throughout the summer. In late May, the Serbs launched an offensive to regain the nearly 40 percent of Kosovo under KLA control.[126] In the process, they drove tens of thousands of Kosovar Albanians from their villages.[127] By early June NATO was considering military options while, on the diplomatic front, the Clinton administration worked through the Contact Group to demand a cease-fire, a Serbian troop withdrawal, and a Kosovar Albanian guarantee to abandon the use of terrorism.[128] By late June, the Serbian offensive had stalled, and the KLA began to step up attacks. Defeats in mid-July, however, forced the KLA back into hiding and a return to guerilla tactics.[129] The Serbs expanded their operations into a broader offensive, displacing hundreds of thousands of Kosovar Albanians.[130]

Prior to the violent opposition of the KLA, Kosovar Albanians had for years peacefully resisted, led by the Democratic Alliance of Kosovo (LDK) and their leader, the intellectual pacifist Ibrahim Rugova.[131] When the United States sought diplomatic contact with Kosovar Albanians it engaged the LDK, and, when Holbrooke met with Rugova in May 1998, he agreed to attend talks in Belgrade.[132] By contrast, the KLA had only recently formed as a secretive insurgent group founded by men lacking in political experience, drawn primarily from the rural villages outside Pristina in western and southeastern Kosovo.[133] For his part, Rugova refused to acknowledge the KLA. By early summer, however, the KLA could no longer be ignored. Holbrooke made at least one attempt to contact their leadership during a visit to Kosovo but was rebuffed.[134]

By late July, however, the Serbian offensive had driven the KLA underground. High desertion rates had reduced their numbers to fewer than 300, rendering them seemingly inconsequential on the military and political front.[135] Holbrooke, therefore, entered talks with Serbia and the LDK having made no contact with the KLA. This structural error of precluding from negotiations what would later turn out to be a key actor in the peace process would eventually cause the Holbrooke agreement to collapse. Despite their exclusion from the talks, the KLA benefited the most from the negotiated Serbian withdrawal, as they seized the opportunity to reorganize, train, and rearm.[136]

The Holbrooke Agreement, October 1998

In September 1998, Russian President Boris Yeltsin and American President Bill Clinton reacted to the humanitarian crisis with a joint statement calling for an immediate cease-fire in Kosovo and the commencement of negotiations. By the end of the month, the UN Security Council had incorporated these demands into a resolution condemning the violence in Kosovo but stopped short of authorizing the use of force by NATO.[137] Russian officials, however, made it known that, although Russia would veto another resolution to authorize force, it would not interfere if NATO elected to conduct air strikes without a UN mandate.[138] Confident that air strikes would not widen the conflict, NATO approved Operation Allied Force (OAF), a limited air operation against Serbian command and control (C2) and military facilities. With the now credible threat of NATO air strikes in hand, Holbrooke returned to Belgrade and brokered a deal with Milosevic on October 12, 1998. Milosevic agreed to reduce Serbian military and police presence in Kosovo, to allow up to 2,000 Organization for Security and Co-operation in Europe (OSCE) monitors to enter Kosovo, and to begin serious negotiations with the Kosovar Albanians.

This interaction between the United States and Serbia leading to the Holbrooke agreement is consistent with the expectations of the asymmetric coercion model. Because of its large advantage in military power, the United States had the latitude to escalate the conflict into a crisis, something it chose to do following the killings of the Jashari clan. The Clinton administration then chose a coercive strategy over brute force. The expected diplomatic and military costs of a ground invasion were much higher than the costs of a limited air campaign and, with nonvital national security interests at stake, the potential benefits from an invasion were low. Clinton therefore adopted a coercive strategy that, at the time, appeared to be only moderate demands that Serbia change its policies regarding Kosovar Albanians. The United States backed up its demands with the threat of NATO air strikes and signaled the credibility of its threat by diplomatic engagement with Russia and the deployment of additional combat aircraft to the region. When Milosevic accepted the Holbrooke agreement, coercive diplomacy succeeded, at least in the short run. A coercion range existed wherein the Americans preferred a negotiated settlement to war and Serbia preferred concession over resistance to U.S. demands and the risk of NATO air strikes. Milosevic likely acceded to these demands because conceding did not threaten the domestic sovereignty of Serbia or the survival of his regime. He miscalculated that Serbia would retain control of Kosovo as he, and Holbrooke, misjudged the relevance of the KLA to the enforcement of the agreement.

Short-Lived Peace: Kosovo Winter of 1998

Holbrooke's diplomatic triumph in October proved short lived, however, as the fatal flaw in the agreement soon surfaced. As Serb forces withdrew from Kosovo, a rejuvenated and defiant KLA stepped from the shadows to occupy the checkpoints and villages they had previously held in May.[139] The militants' actions prompted Serbia to reverse course and redeploy its forces.

Milosevic learned from this experience that, without the compliance of the KLA, withdrawing Serbian forces from Kosovo was tantamount to conceding domestic sovereignty over Kosovo. This he refused to do, even under the threat of NATO air strikes. Furthermore, late December's Desert Fox air strikes against Iraq demonstrated to Milosevic that Saddam Hussein's regime could survive the restricted air strikes that Serbia now faced. Coercive diplomacy now failed as the limited threat of force no longer matched the high level of demands being made.

In December and January, violence returned to Kosovo in a series of tit-for-tat reprisal attacks between the KLA and Serbs. On January 14, 1999, in response to a KLA attack that left three policemen dead, Serb forces entered the village of Račak, killing forty-five men. William Walker, chief of the OSCE monitors, arrived the next day and placed responsibility for the attack squarely on the Serbs.[140]

Holbrooke's October agreement had provided only temporary relief. Although it did allow Kosovar Albanian refugees to return to their homes before the winter, it did not create a sustainable solution to quell the violence.[141] The initial stage of the crisis ended in January with the massacre at Račak, which signaled to Clinton that his foreign policy was no longer tenable.[142]

The Crisis Escalates: January to Late April 1999

The international attention created by William Walker's condemnation of the Serbs led to a change in White House policy. Albright now convinced National Security Advisor Sandy Berger to support a more aggressive position.[143] On January 19, the National Security Council Principals Committee agreed to an ultimatum for the removal of Serbian forces from Kosovo and the insertion of NATO troops as peacekeepers.[144] Still, Cohen and Shelton stood fast in refusing to consider U.S. ground troops for combat operations.

At an international conference at Rambouillet, France, in February, a final attempt at coercive diplomacy produced the "Interim Agreement for Peace and Self-Government in Kosovo." This called for an aggressive timetable for the withdrawal of Serbian forces (except for border guards), the deployment of a NATO implementation force, the establishment of a democratic Kosovo government, and an international meeting to be held in three years' time for a final settlement.[145] Unlike negotiations the previous fall, Milosevic no longer showed any interest in a compromise. Four changes help to explain this reticence. First, the Serbian population disapproved of the Rambouillet proposal. In opinion polling conducted during the talks, the Serbian population overwhelmingly opposed foreign intervention in Kosovo.[146] Second, Milosevic's political coalition had moved to the right to include the Serbian Radical Party, and a purge of moderates further removed those from his government and military who might argue for compromise.[147] Third, in December, Milosevic had witnessed Saddam's regime survive four days of bombardment, and he sent Serbian defense specialists to Baghdad to glean lessons on how to withstand U.S. air strikes.[148] Finally, the demands being made at Rambouillet were significantly greater than those Milosevic thought he had agreed to back in October.

The United States now demanded Serbia cede control of Kosovo. Equally important, the White House had mismatched its threats to its demands. Although the demands had been significantly raised, the threat of limited air strikes had remained unchanged. Milosevic had already signaled that he would not give up Kosovo when he redeployed Serbian forces in November in response to the reemergence of the KLA. For the United States to gain territorial concessions through coercive measures would require a significant increase in the level of force being threatened. Instead, the Clinton administration maintained its previous threat of limited air strikes. Three factors explain the inability/unwillingness of the United States to make threats commensurate with demands. First, the United States did not have vital national security interests at stake. The Clinton administration was primarily concerned with preventing another humanitarian disaster similar to what had occurred in Bosnia. This effectively capped the level of force they were willing to threaten. Cohen and Shelton adamantly opposed the deployment of ground troops for any role other than peacekeeping. Clinton likewise had little appetite for a ground war in Kosovo. He would later make public the fact that he had already ruled out such an option.[149] This was a strategic blunder as it reduced uncertainty as to whether NATO might invade. Absent this tipping of the hand, however, Milosevic still had reason to be confident, given the transparency of NATO's operational planning, that a ground invasion of Kosovo was unlikely.

Second, a rift between the State and Defense departments resulted in the use of separate criteria for evaluating the level of demands the Clinton administration would make and the threat of force they would use to back up those demands. The Račak massacre proved a turning point for Clinton's foreign policy with regards to Kosovo. Albright pinpointed Milosevic as the problem and argued for getting Serb police and military out of Kosovo, NATO troops in, and a return to Kosovo autonomy.[150] Cohen and Shelton, who argued against threatening military force over nonvital interests, had previously checked her viewpoint. After Račak, however, these two were effectively silenced as Berger swung his support to Albright. Even so, Cohen and Shelton held firm on not sending in ground troops. Clinton eventually agreed to a foreign policy that fundamentally mismatched demands and threats. Why did he agree to such a flawed plan?

The primary reason for the administration's miscalculation appears to be a misperception of Milosevic's resolve. Clinton, Albright, and Holbrooke had all faced Milosevic before in Bosnia and again in October 1998 when he backed down under the threat of force. They had come to believe that the threat of air power, or at most three nights of strikes, would compel him yet again. They failed, however, to properly consider Milosevic's interests in Kosovo. Kosovo was not Bosnia. Kosovo was the historic birthplace of Serbia, the retention of which had served as the platform for Milosevic's ascension to power.

Serbia was not alone at Rambouillet in rejecting the peace plan. So, too, did the KLA-led Kosovar Albanian delegation. They sought a referendum for Kosovo independence, a measure opposed by both the Serbs and the Russians. Instead of a referendum, the United States inserted an amendment, designating that in "three years after the entry into force of the Agreement, an international meeting [would] be convened to determine a mechanism for a final settlement for Kosovo."[151] It took Albright threatening to take NATO air strikes off the table before the KLA relented.[152] Even then, its representatives would not sign without first returning to Kosovo to explain the agreement to their commanders.[153] Once satisfied that the KLA leadership concurred, the delegates returned and endorsed the agreement on March 18, 1999. With signatures in hand, Holbrooke returned to Belgrade, where he met for the last time with an intransigent Milosevic. On March 24, 1999, NATO commenced air strikes.

Allied Force: NATO's Air Campaign

The initial phase of Operation Allied Force (OAF) took aim at fifty NATO-approved targets: Serbian integrated air defense sites (IADS), command and control facilities, airfields, military and police barracks, electric power facilities near Pristina, and two "dual-use" factories.[154] Given the combat aircraft and cruise missiles available to NATO, this target set should have required only three nights of attacks.

NATO commenced operations with 214 combat aircraft, roughly half of which were U.S. strike aircraft. In addition, NATO employed conventional cruise missiles launched from B-52s, U.S. Navy surface ships, and U.S. Navy and British HMS submarines.[155] Though other NATO countries also provided tactical aircraft, U.S. aircrews received the highest priority targets.

NATO aircraft faced a nontrivial threat from the Serbian integrated air defense system. This professionally trained and robust IADS consisted of older-generation Soviet radar-guided missiles, vehicle-mounted infrared-guided missiles, AAA (antiaircraft artillery), and shoulder-launched, infrared-guided MANPADS (man portable air defense systems).[156] In addition, the Serbian Air Force flew a squadron of modern M-29 fighter interceptors, along with several squadrons of older MiG-21s.[157] NATO would effectively neutralize these defenses by jamming and attacking Serbian radars, flying at medium altitude well above MANPADs and AAA effective ranges, and by employing Combat Air Patrols (CAPs) to protect strike packages from Serbian fighters. Such defensive tactics came at a cost to NATO, however, as the number of air strikes would be restricted by the availability of electronic jamming aircraft and medium altitude tactics would make it more difficult to identify and attack Serbian forces.

The first phase of OAF from March 24 to 27, 1999, focused on suppressing Serbia's air defenses and targeting Serbian military facilities. On the first night, NATO launched cruise missiles against Serbia's IADS and airfields and Kosovo's electrical power plant in Pristina. In addition, NATO jets conducted air strikes on radar and SAM sites, military airfields, command and control nodes, military barracks, and munitions storage areas. NATO fighters also downed three MiG-29s.[158] Although NATO flew even more strike missions the second night, poor weather on the third night extended the strikes into a fourth night.

The first four nights of air strikes, however, did not convince Milosevic to change his mind, nor did it deter Serbian forces from commencing large-scale ethnic cleansing operations against the Kosovar Albanians. Serbia responded in two ways. First, the air strikes galvanized the population to "rally round the flag" in support of Milosevic. At night in Belgrade and other cities, as air raid sirens blared, thousands of Serbs defiantly congregated in the streets and on bridges.[159] A demonstration as to the degree of solidarity was evidenced on March 28, when a rock concert commemorated the ten-year anniversary of the constitutional changes that had stripped Kosovo of its autonomy. Many of the bands that played and the Serbian students who now protested NATO air strikes had been the same ones to protest Milosevic's falsification of the 1996 election results.[160]

Second, within Kosovo, Serbian ground forces now punished the KLA by targeting the Kosovar Albanian population.[161] Within days, tens of thousands poured across the border into makeshift refugee camps in Macedonia, Albania, and Montenegro. After the initial phase of NATO bombings, the numbers of refugees would swell to hundreds of thousands as the Serbs moved to empty Pristina and towns throughout Kosovo.[162] By mid-April, Serbian forces would again be in control of Kosovo whereas the KLA had fled along with the refugees.

Efforts at coercive diplomacy by only threatening force failed once NATO commenced OAF. These restricted air strikes did not change Milosevic's mind, however, triggering instead an escalation in military operations against the KLA. This reaction by Serbia, in turn, fundamentally altered the nature of the conflict. The ethnic cleansing of Kosovar Albanians proved to be a tactical success but a strategic blunder. Rather than enduring only a few days of attacks, as Saddam Hussein had with Desert Fox, Milosevic now confronted daily air strikes with no end in sight.

Extension of Allied Force, March 28–April 24, 1999

The United States responded to the large-scale humanitarian crisis by expanding its target list to include the Serbian fielded forces in Kosovo. These air strikes, however, proved to be ineffective due to limited contingency planning for such operations, restrictions to the rules of engagement, lack of friendly ground troops to facilitate strikes, poor weather, and tactical adaptation by Serbian forces.[163] After a month of bombing in Kosovo, NATO had neither stopped the violence nor seriously weakened the Serbs' military capability.[164]

In addition to targeting fielded forces, on April 3 NATO conducted a small number of strikes on leadership targets in Serbia.[165] Two security headquarters housed in office buildings were struck in downtown Belgrade, followed over the next two weeks by strikes on police and military headquarter buildings, telephone exchanges, TV and radio stations and towers, and dual-use factories and oil refineries.[166] NATO also struck transportation and other infrastructure targets. By April 21, the last bridge over the Danube had been dropped and Belgrade's main water supply destroyed. NATO also began targeting Milosevic and his political supporters by bombing the office building housing the headquarters of Milosevic's Socialist Party of Serbia (SPS) and his wife Mirjana Markovic's Yugoslavia Left (JUL) party. Bombs struck the master bedroom of Milosevic's official residence the next day.[167]

Even with attacks directed at Milosevic personally and an increased number of strikes against Serbian fielded forces, it was clear by late April that the current level of strikes would not coerce Milosevic. As the White House hosted the NATO Summit to celebrate its fiftieth anniversary, U.S. interests had expanded beyond that of international humanitarian concerns to include its own security interests in a viable NATO and the prestige of the United States and of Clinton's presidency. As this second coercive stage concluded, the United States found itself frustrated with its inability to compel Serbia, though this would soon change.

THE FINAL ACT, APRIL 24–JUNE 9: FROM WASHINGTON SUMMIT TO PEACE AGREEMENT

On April 24, as NATO leaders arrived in Washington, D.C., two critical developments altered the dynamics of the crisis. First, Yeltsin phoned Clinton and offered to pressure Milosevic into a peace agreement.[168] The United States agreed to meet with Yeltsin's new personal envoy to the Balkans, Viktor S. Chernomyrdin, to negotiate a mutually acceptable proposal to present to Milosevic.[169] Second, Clinton and other leaders arrived at a consensus that the reputation and credibility of NATO were now at stake. With U.S. security interests at play, the United States and its NATO allies emerged from the Washington Summit with a newfound resolve for victory in Kosovo.[170] Economically, they imposed additional sanctions on Serbia (and Montenegro), including a ban on oil sales, a prohibition on travel, and the freezing of financial accounts.[171] Militarily, NATO leaders approved an escalation of attacks on civilian and leadership targets just as additional combat aircraft arrived in theater.[172]

As the intensity of air strikes escalated, however, so, too did the likelihood of mishaps. There soon followed two incidents that diluted NATO's efforts.

First, on May 5 a U.S. Army Apache attack helicopter crashed, killing two soldiers. This was the second such incident for the Army's Task Force Hawk during mission rehearsals in the high mountains along the Kosovo–Albanian border.[173] The Apache deployment was intended to place additional pressure on Serbian forces in Kosovo. It instead turned into a public debacle as the Pentagon intervened, blocking Supreme Allied Commander Europe (SACEUR), General Wesley Clark's bid to employ the task force.[174] Second, and more important, on May 8 a B-2 bombed the Chinese Embassy in Belgrade, killing three Chinese personnel. This resulted in the suspension of all strikes around Belgrade for three days and sidetracked American–Russian negotiations for nearly two weeks.[175]

Despite the diplomatic fallout over the Chinese Embassy bombing, overall NATO still increased its operations tempo as additional aircraft arrived and the weather improved. In May, it nearly doubled the targets attacked, tripled the weapons expended, and increased by half again the number of missions flown.[176] Although previous air strikes against bridges, railways, and communication sites had inconvenienced the Serbian population, attacks on Serbia's electric grid now posed a more serious threat. By mid-May, Serbia had up to 85 percent of its electrical power interrupted, which, in addition to the direct impact on daily life, also had the ripple effect of causing disruptions in water supplies to cities throughout the country.[177]

Serbs were also beginning to feel increased economic pressure as their already weakened economy stagnated and unemployment began to rise. Air strikes on factories and transportation networks and the disruption of the electrical power supply combined with the economic sanctions on oil imports and the freezing of financial assets resulted in a 30 percent drop in Serbia's already anemic economic output.[178] Serbian officials reported that the bombing of factories alone had put more than half a million additional workers out of work, with a total loss of two million jobs as a direct result of the air campaign by June.[179]

Two months of bombings and the inconveniences of power outages, lack of water, and shortages of imported goods began to take their toll on a weary Serbian population as their exuberance for war waned. The daily rock concerts sponsored by Mirjana Markovic's JUL party petered out as the tens of thousands who first demonstrated dwindled down to a few hundred.[180] In the south, women began protesting against the Serbian Army for having their husbands and sons serve in Kosovo; once their men had been granted leave, they protested their return to duty.[181] Though there was growing displeasure, the Serb population, for the most part, still blamed the United States rather than Milosevic for their predicament. He thus managed to avoid the large anti-Milosevic demonstrations he had encountered following the 1996 elections.

The first visible sign of domestic political opposition came in late April from Deputy Prime Minister Yuk Draskovic, the leader of the nationalist SPO party who had joined Milosevic's coalition the previous year. Draskovic publicly accused Milosevic and Markovic of using the war for political gain and called for a negotiated settlement with NATO. His comments succeeded only in getting him removed from office.[182] By mid-May, however, other Serbian officials began publicly advocating for an end to the war. In fact, the idea of allowing U.S. ground forces into Kosovo as peacekeepers was being openly debated within Milosevic's regime.[183] By the end of May, small and sporadic anti-Milosevic protests had sprung up in some of the more severely bombed cities, and several opposition party leaders had begun to openly criticize Milosevic.[184]

Russian and U.S. Diplomacy

Back in mid-April, Boris Yeltsin had become dissatisfied with Russia's foreign policy efforts to bring about an end to the Kosovo War and replaced his Prime Minister, Yevgeny Primokov, with Viktor Chernomyrdin.[185] Following Yeltsin and Clinton's phone call during the Washington Summit, Chernomyrdin traveled to Washington on May 4 to deliver a letter. In it, Yeltsin proposed a cease-fire and requested a U.S. envoy be named to assist in Russian diplomatic efforts.[186] Clinton chose Strobe Talbott, who was well acquainted with the Kremlin, having previously worked on the dissolution of the Soviet Union as special advisor to the secretary of state.

Chernomyrdin and Talbott's efforts led to a joint endorsement of a seven-point peace plan released at a G8 Foreign Minister meeting on May 6.[187] The plan modified the Rambouillet Accords to allow the UN Security Council to determine the make-up of peacekeepers and to affirm "the principles of sovereignty and territorial integrity" for Serbia.[188] Though diplomatic efforts stalled for nearly two weeks in the wake of the Chinese embassy bombing, by May 20 Chernomyrdin presented the G8 proposal to Milosevic.[189] The key sticking point was whether any Serbian troops would remain in Kosovo and the composition of foreign peacekeeping troops. On June 3, Milosevic agreed to a modified G8 proposal that substituted UN for NATO peacekeepers, thus placing the future of Kosovo in the hands of the Security Council.[190] Military negotiations for implementing the peace agreement took place along the Kosovo–Macedonian border from June 5 to 9.[191] Shortly thereafter, NATO halted its bombing while Serbian forces withdrew. A UN Security Council Resolution passed on June 10 acknowledged an end to the fighting and the deployment of an international peacekeeping force.[192] Russian and NATO ground forces then crossed the border into Kosovo on June 12, 1999.[193]

Coercion Success: Why Milosevic Gave up When He Did

What caused Milosevic to finally accept the G8 peace proposal after two months of NATO bombing has been the subject of much debate.[194] The reasons he changed his mind when he did are important for an understanding of what can and cannot be achieved by coercion. As discussed in Chapter 2, a successful coercive strategy makes demands that are backed up by a commensurate and credible threat of force. At Rambouillet, however, the United States had raised its demands from a moderate policy change over how the Kosovar Albanians should be treated to a much more serious demand that Serbia concede domestic sovereignty of Kosovo. The survival explanation for coercion failure introduced in Chapter 3 expects that, under such conditions, coercion should fail. Demands for homeland territory should be perceived as a threat to Serbia's survival. As a result, Serbia should have resisted, even though its chances of victory against the United States would have been small. And yet coercion succeeded.

The case of Kosovo is unique in that it is the sole case in which the United States has succeeded in coercing a weak state to forfeit homeland territory. A further examination of Kosovo provides insights into why Milosevic conceded. Though Serbians considered Kosovo their historic birthplace, few Serbians resided there. Kosovo also had a history of autonomy, having been a Yugoslavian province from 1974 until 1989. Also, given its small economic output, its tiny landmass, and its insignificance geopolitically, Kosovo was of marginal concern to Serbians. Serbia, in fact, could survive without Kosovo, and, though it was initially willing to fight for it, in the end its loss did not spell the end to the state of Serbia.

Though Serbia resisted U.S. efforts at coercive diplomacy at Rambouillet and continued to resist when faced with air strikes in late March and April, the United States eventually compelled Milosevic to accept the G8 proposal. Four factors explain why he made this decision when he did: First, Milosevic's strategy was no longer working; second, by continuing the war he was losing the political support of the population and elite; third, time was against him as he faced a credible, imminent threat of even more costly strikes to Serbia's economic infrastructure; and fourth, the G8 proposal provided him a face-saving means for making concessions.

The first contributor to Milosevic's concession was the mounting evidence that his strategy no longer had much chance of succeeding. His strategy can be usefully disaggregated into two military and two political components. The first military objective seized control of Kosovo and presented NATO with a fait accompli. Prior to the first air strikes of OAF, Serbia had already deployed additional troops and equipment to carry out a counterinsurgency operation. Protected from NATO air strikes by cloud cover, Serbian forces pressed forward to evict the Kosovar Albanians from Kosovo and, in so doing, remove the source of KLA support. From a tactical perspective, these operations were highly effective. The KLA was either forced into hiding or departed Kosovo with the other refugees and, even when the KLA was able to reorganize and attempt an offensive in late May, their forces were easily repulsed.[195] From a strategic standpoint, however, this proved to be a critical mistake. Milosevic may well have expected the flow of refugees into neighboring countries to provide him the bargaining power to leverage against NATO for a better deal in exchange for allowing the refugees to return home.[196] The CNN images of thousands of suffering Kosovar Albanian refugees instead lent credence to claims that Serbia, and not the United States, was at fault in Kosovo. This reinforced NATO's justification for continuing and escalating air strikes.

The second military component to Serbia's strategy was to inflict significant combat losses on NATO aircraft and aircrew, making it either too costly for NATO to continue air operations or, at a minimum, creating tension among NATO countries that might cause a fissure in the alliance. The Serbian IADS engaged NATO jets from the opening strikes until the final days of OAF, but their efforts were largely neutralized. Serbia succeeded in downing only two aircraft, but U.S. rescue assets recovered both pilots, and Serbia was never able to generate losses significant or consistent enough to cause NATO to question the risk level of its air campaign.

Along with its military efforts to attrit NATO's forces, Serbia also employed the political strategy of attempting to fracture the alliance by exploiting civilian deaths caused by the bombings. The United States was fully aware of the negative effect errant bombs would have on NATO's fragile consensus.

Collateral damage estimates and risk calculations were therefore incorporated not only into the selection of targets but also into the timing of attacks. For example, the two factories hit on the first night of OAF were, in part, targeted because they were inactive, which significantly reduced the risk of civilian casualties.

The first major collateral damage incident took place on April 13 when a U.S. jet struck and then reattacked a bridge as a train crossed it, killing ten passengers.[197] This event was trumped the next day, however, by U.S. pilots misidentifying tractors in a column of Kosovar Albanian refugees as military vehicles. The ensuing attack killed seventy-four.[198] Although these back-to-back blunders placed NATO on the defensive diplomatically, the refugee attack proved counterproductive to the Serbian media campaign. The Serbs seized the initiative, bussing international journalists to the scene to document the grisly event. Enroute, however, the reporters observed the torched homes and vacated villages of Kosovar Albanians. They then interviewed survivors, who provided firsthand accounts not only of the aerial bombing but also of Serb troops and police forcing them from their homes at gunpoint.[199] After some initial faulty reports, NATO admitted responsibility for the refugee attack and tightened its rules of engagement to prevent a recurrence. In the end, however, the overriding message was that, although NATO's attacks on the Kosovar Albanians were an accident, the actions of the Serbs clearly were not.

No further major collateral damage incident occurred until NATO began escalating operations following the Washington Summit. The bombing of the Chinese embassy on May 8 was not the most deadly, but it was by far the most politically harmful collateral damage event of the war. Not only did it sidetrack U.S.–Russian negotiations, but it also threatened U.S–Chinese relations. Nonetheless, this and subsequent collateral damage events signaled NATO's newfound resolve to see the war through despite the additional risk.[200] Serbia was never able to fracture the NATO alliance over the issue of collateral damage. Any opportunity Serbia may have had in the early stages of the crisis was more than offset by the images of Kosovar Albanians driven from their homes.

The second political component to the Serbian strategy lay in Russian intervention. At the commencement of Allied Force, this appeared promising as the Kremlin decried Western aggression while angry mobs demonstrated outside the U.S. embassy in Moscow. Primakov, when notified of the NATO air strikes while aboard his Washington-bound jet, dramatically ordered the plane be turned around over the Atlantic.[201] Though angered by NATO's actions, Yeltsin depended on American economic assistance and could not afford to be drawn into the conflict militarily. Yeltsin dispatched Primakov to Belgrade with orders to bring an end to the crisis. Despite Primakov's efforts, Milosevic refused to negotiate unless NATO first ceased bombing, a precondition the White House would not consider.[202]

Two weeks into the air war, Yeltsin, who faced fierce domestic opposition and possible impeachment hearings in the Duma, was clearly frustrated over the hostilities and the inability of Primakov to bring Milosevic to the table.[203] On April 14, the Kremlin announced that Primakov had been replaced by Viktor Chernomyrdin. This marked a policy shift for Russia. Yeltsin now pushed for a cease-fire on terms much closer to those preferred by the United States. Evidenced by Yeltsin's phone call to Clinton on April 24 and by Chernomyrdin's negotiations with Strobe Talbott through the end of May, this new policy removed the final underpinning to Milosevic's strategy. Chernomyrdin traveled to Belgrade on June 3 to deliver the G8 proposal to a dejected Milosevic, who reluctantly acceded.

By June it was clear to Milosevic that his strategy of resisting NATO had not worked. He had not been able to stop the bombings by fracturing the NATO alliance. Attempts at downing a large number of aircraft had failed, as had his touting of civilian casualties. Serbia's military operation had succeeded in evicting the majority of Kosovar Albanians from their homes, but this had not gained Milosevic the additional bargaining power he had anticipated, serving instead to neutralize any international sympathy Serbia may have garnered as the victim of NATO's military action. Finally, Yeltsin's decision to work with Clinton dashed Milosevic's last hope of Russian intervention.

The failure of Milosevic's strategy is a necessary, though insufficient, explanation for why he chose to concede to the G8 proposal when he did. Just because his strategy had not worked did not mean he had to accept the peace plan, unless there were also significant costs being imposed on him for continuing to resist. The second reason he gave up in early June was the political costs incurred as support for the war waned among the Serbian population and the political elite. Indeed, general popularity and the backing of the old communist establishment had been key to Milosevic's rise to power. Although Milosevic was deft at political maneuvering, including electoral fraud to remain in power, there were limits to how much he could manipulate Serbia's electoral system. This was witnessed back in the 1996 election when only after months of widespread protest did he finally relent and accept the official results. Though Serbia was clearly not a democracy, the vote of the Serbian population still mattered. As Milosevic's popularity decreased throughout the 1990s so, too, did his power base and freedom of action.

The Serbian population's eagerness to resist NATO, demonstrated by the hundreds of thousands who protested in late March, had by May been replaced by war weariness. The rock concerts, once cheering on Milosevic, now devolved into a small but growing number of protests against the war, as widespread power and water outages, rising prices, and unemployment had sapped popular support. A final piece of ex post evidence of their sentiments was the general sense of relief and the lack of protests that greeted Milosevic's announcement of a peace agreement.

The political elite had, like the population, initially either supported the war or at least felt constrained against speaking openly against the war or Milosevic. By mid-May, however, opposition leaders, as well as those within Milosevic's own party, felt emboldened to publicly call for concessions. Defiance against NATO had now shifted to acceptance of an agreement to stop the bombing, even though this meant foreign troops in Kosovo.

Milosevic valued Kosovo and was willing to go to war with NATO over it. Remaining in power, however, was something he valued even more, especially now that his strategy had not worked. To remain in power would require the support of the Serbian population and the elite, both of which were turning against the war. Although vulnerable at the ballot box, unlike Saddam Hussein who had been threatened by the Shi'ites, Milosevic did not face violent internal opposition. There were no domestic groups within Serbia with the capacity to violently overthrow his regime. Back in 1990, Milosevic had sealed an alliance with his government and the Yugoslavian Army. This successfully subjugated the military to civilian control, thus lowering the chances not only of a military coup but also of the formation of armed opposition.[204] As a result, Milosevic did not have to worry that conceding to the G8 proposal would lead to an insurgency.

A third explanation for Milosevic's decision was his resigned sense that a delay would only lead to further damage to Serbia's infrastructure and economy. The ever-improving weather, increased number of NATO combat aircraft, and the shift to strikes on targets throughout Serbia resulted in the expected cost for rebuilding, already estimated in the billions, to climb even higher.[205] Despite previous attacks on bridges, factories, oil facilities, and the power grid, NATO had demonstrated a good deal of restraint, given the infrastructure and economic targets still available to attack. Economically, Serbia stood to lose a great deal should Milosevic refuse the G8 proposal. And though the Serbian people had suffered economically through most of Milosevic's tenure in office, this did not make him immune to their plight. Indeed, during the Bosnian Civil War he had held the lifting of economic sanctions as his highest political objective. With his strategy derailed and the population and elite turning against him, delay would only increase costs to Serbia and decrease the chance he could win future elections.

Rather than the anticipation of further punishing air attacks against Serbia's economy, an alternative explanation for Milosevic's decision to concede to the G8 proposal lay in the mounting threat of a NATO ground invasion.[206] Clearly the potential for a NATO ground invasion had increased following the Washington Summit. British Prime Minister Tony Blair pressured the White House to consider the issue, and on May 19 Clinton relented by announcing that he would not rule out a ground option, though he did add that the alliance "ought to stay with the strategy that we have and work it through to the end."[207] Two days later, the Clinton administration called for NATO to deploy 50,000 troops along the border. Though the purpose for the deployment given by the White House was the preparation for a peacekeeping mission, the timing of the move added weight to the likelihood of invasion.[208] On June 2, one day before Milosevic conceded, Clinton met with the Joint Chiefs of Staff for the first discussion of a ground invasion. To have the requisite 150,000 troops in place by mid-August in order to avoid a winter ground war, a decision to invade needed to be made by mid-June.[209]

The shortcoming of this counterargument is that the key actions that would have made the threat of a ground war credible had not yet been carried out. Even though Blair lobbied hard for it, Clinton was still deliberating, and a reluctant Congress had not yet approved the large-scale deployment of ground troops necessary for a ground war. The forces that were in place were not sufficient to mount offensive action, and it would take months to remedy this shortcoming. Likewise, the ground threat from a KLA offensive had not materialized. The Mount Pastrik operation in western Kosovo that took place at the end of May failed miserably.[210] The Serbian military remained entrenched overlooking the major arteries into western and southern Kosovo, and the expected casualties to U.S. ground troops, even with NATO's air supremacy, would have been significant. Overall, the threat of a ground invasion does not explain Milosevic's willingness to accept the G8 proposal on June 3. The imminent threat of additional economic losses from further air strikes better explains the timing of his actions.

A final reason for Milosevic's willingness to concede to the G8 proposal was its inclusion of face-saving measures absent from the Rambouillet demands. This allowed Milosevic to make the claim to the Serbian people that he had not surrendered Kosovo to the United States. The deployment of Russian peacekeeping troops under the auspices of the United Nations was a key U.S. compromise. This provided Russian influence both on the Security Council and in Kosovo to ostensibly protect Serbia's interests. Further, the G8 proposal removed all reference to a future referendum on the question of Kosovo independence. These two measures provided the cornerstone to Milosevic's address to the Serbian people on June 10 after the peace agreement had been finalized.[211] Over two months of resistance would appear to have brought about U.S. compromise in the G8 proposal, thus providing a platform, albeit shaky, for Milosevic to make an argument to the Serbs that enduring seventy-eight days of air strikes had been justified and that he had not, in fact, surrendered Kosovo.

The Asymmetric Coercion Model and Alternative Explanations for War

Kosovo is an important case for examining the limits of what can be achieved by asymmetric interstate coercion. The United States would not accommodate Serbia in its treatment of the Kosovar Albanians, choosing instead to escalate the conflict into a crisis. With only nonvital national security interests on the line, however, Clinton ruled out a brute force strategy in favor of coercion. In the fall of 1998, the United States made moderate demands of Serbia to change its policies regarding Kosovar Albanians, but, by the spring of 1999, its demands had increased to Serbia ceding domestic sovereignty over Kosovo. Coercive diplomacy failed, as threats to back up these demands were limited to only three nights of restricted air strikes. Over time, however, the cooperation by Russia, coupled with the increase in NATO combat operations, signaled a significant increase in the threat of force. By May, the continuation of the air war threatened the entire Serbian economy. With the Serbian population and elite turning against the war, Milosevic finally conceded, but not before he extracted concessions in an effort to lower his audience costs. A coercion range finally materialized in June, wherein the United States preferred the G8's proposed outcome to a brute force war and Serbia preferred accepting the proposal to continued resistance.

The asymmetric coercion model and coercion range help to explain the interaction between the United States and Serbia that ultimately led to a successful coercive outcome. By contrast, as previously discussed, compelling Serbia to concede its homeland territory was an accomplishment not anticipated by the survival explanation for coercion failure, nor was it expected by other conventional explanations for war. Misperception, miscalculation, and uncertainty, as with Bosnia, again help to explain why coercive diplomacy failed and why it took so long for the United States to achieve a successful coercive outcome. Initially, both the White House and Milosevic misperceived the demands being made in October 1998. Only with the withdrawal of Serbian troops and reemergence of the KLA did it become clear to Milosevic that the demands being made were for homeland territory. It is unclear, however, when it became apparent to the Clinton administration that the demands they had made were greater than simple policy change in Kosovo. Biases regarding Milosevic's resolve and internal disagreements in his own administration caused Clinton to miscalculate that three nights of air strikes would be sufficient to convince Milosevic to accept the Rambouillet Accords. Though misperception, miscalculation, and uncertainty help to explain why coercive diplomacy failed, they fail to provide useful insight into why coercion ultimately succeeded. The best that can be offered is a tautological argument that misperception, miscalculation, and uncertainty must have decreased over time to the point where a bargained outcome could be reached.

As with the Bosnian Civil War, commitment problems also do not appear to have hampered Serbia in acquiescing. The logic of commitment problems is that a shift in the balance of power or the revelation that the target lacks resolve by conceding introduces an incentive for the challenger to make further demands. Russia's involvement in negotiations and the assurance of its peacekeeping troops, however, reduced the likelihood of this occurring as it introduced additional diplomatic costs for the United States to renege on its promises. In addition, because other NATO countries were involved in operations, it was unlikely that the United States could have convinced the other eighteen NATO nations to further increase demands. The Americans had again effectively reduced potential commitment concerns by involving a third party great power and conducting military operations within a coalition.

CONCLUSION

This chapter examined two important cases of coercion in which the United States adopted strategies that employed air power and sanctions but intentionally did not risk ground forces, as vital national security interests were not at stake. Bosnia and Kosovo are lengthy crises for which the United States eventually succeeded in achieving its core demands but not before it had placed its reputation and prestige on the line. The United States risked all-out war and expended a considerable military effort for relatively meager returns. Because of the delay in applying sufficient coercive pressure to bring an end to the two crises there resulted a considerable amount of bloodshed, ethnic cleansing, and severe economic losses on the part of Bosnians, Serbians, and Kosovar Albanians. Still, despite these tragic events and the failure of coercive diplomacy, the United States did eventually compel an end to the Bosnian Civil War and wrested Kosovo from Serbian control.

The survival explanation for coercion failure correctly expected the concessions made by Milosevic and the Bosnian Serbs to end the Bosnian Civil War. With Kosovo, however, the survival hypothesis incorrectly anticipated demands for homeland territory to be rejected as a threat to Serbia's survival and a source of significant audience costs for Milosevic. He conceded, however, despite the loss of domestic sovereignty over Kosovo, because he believed further resistance to be futile and threatened the economic viability of Serbia and, by extension, his grasp on power.

The next chapter continues to evaluate the capability and limitations of coercion in asymmetric interstate crises with the more than two decades long conflict between the United States and Libya from 1981 to 2003.

6 THE UNITED STATES VERSUS LIBYA
El Dorado Canyon, Pan Am Flight 103,
and Weapons of Mass Destruction

THE INTENTION OF THIS BOOK is to provide policy makers and analysts with a set of methodological tools to be able to systematically examine and explain conflicts in which the United States confronts much weaker states. In Chapters 2 and 3 an asymmetric coercion model was developed, a coercion range was defined to explain coercion success, and a target survival hypothesis was introduced to explain why coercion often fails. In Chapters 4 and 5 these analytical devices were employed to examine U.S. asymmetric interstate conflicts against Iraq and Serbia, respectively. Chapter 4 assessed the Gulf and Iraq wars, for which coercion failed because Iraq was ultimately unwilling to make concessions that threatened the survival of the Iraqi state and/or Saddam Hussein's regime. In these cases, the United States was not surprised at the failure of coercion as coercion was but an initial stage of a mixed strategy. In an effort to obtain UN Security Council resolutions to justify the brute force strategies already decided on, the United States first agreed to coercion. This is true even though the White House understood that the foreign policy objectives it had chosen were unlikely to be obtained by coercion alone. Chapter 5 analyzed crises in Bosnia and Kosovo, in which the United States was pitted against the Bosnian Serbs and Serbia, respectively. Unlike the conflict in Iraq, in the Balkans the United States did not perceive vital national security interests to be at stake and therefore chose not to endure the high costs of brute force invasions. As with Iraq, coercive diplomacy again failed to achieve a peaceful outcome, but coercion by air power eventually succeeded, though not before thousands of Bosnians, Serbians, and Kosovar Albanians had died, hundreds of thousands more had been made refugees, and the reputation and prestige of the United States and NATO had been placed on the line.

This chapter continues to examine asymmetric interstate conflicts through the lens of the coercion model, the coercion range, and the survival hypothesis. Three long-term conflicts between the United States and Libya from 1981 to 2003 are useful for expanding our understanding of coercion. In the first case, tensions between the United States and Libya arose over Libyan leader Muammar Qaddafi's support of international terrorism. As the conflict continued, the Reagan administration added regime change as a foreign policy objective. The conflict escalated to a crisis with the Christmas bombings in Rome and Vienna in 1985. The crisis climaxed with the El Dorado Canyon air raid on April 15, 1986, when U.S. jets attacked Qaddafi's headquarters in Tripoli and terrorist training facilities in Benghazi. Qaddafi survived, but he subsequently reduced his anti-U.S. rhetoric, and Libya stopped overtly supporting international terrorist groups to prevent President Reagan from having the justification for further strikes. As a result, the United States was in the long run unable to maintain a credible threat of force and failed to compel Libya to abandon completely its terrorist policies, as evidenced by the tragic explosion of Pan Am Flight 103 over Lockerbie, Scotland, on December 21, 1988.

The second case arose in 1991 once the United States and Britain had obtained forensic evidence linking the Pan Am bombing to the Libyan government. They demanded that Libya cooperate, take responsibility, and extradite two officials to either the United States or Britain to stand trial. This differs notably from the Iraqi and Serbian cases in that U.S. demands decreased over time under President George H. W. Bush, who, unlike his predecessor, no longer pursued Libyan regime change. The level of threats issued by the United States was likewise reduced as American leaders no longer considered military force, relying instead on economic sanctions.

Based on the definition introduced in Chapter 2, this second case is not a crisis because there was not a heightened probability of military hostility accompanied by a time constraint.[1] There are, however, similarities between strategies that threaten economic sanctions and those that threaten military force. A powerful challenger can employ both strategies to punish and thereby compel a weak target. As a result, this case is considered to assess how well the asymmetric coercion model also explains the outcomes of conflicts for which sanctions replace the threat of force.

Given the vulnerability of Libya's depressed economy to international sanctions and the fact that the demands were relatively minor and did not threaten the survival of the Libyan state, Qaddafi was willing to entertain making concessions. He would not, however, accept the humiliation of extraditing his men to stand trial in America or Britain, nor would he take responsibility and pay compensation before a trial had taken place. Qaddafi instead countered by offering to extradite the two officials to The Hague. When the United States refused, a stalemate ensued for six years. In 1997, the situation was finally resolved when President Bill Clinton, with the encouragement of British Prime Minister Tony Blair, agreed to Qaddafi's conditions. By having the United States, and not Libya, make the final compromise, Qaddafi saved face and avoided the audience costs for being compelled to extradite his men.

For the next four years, the United States and Libya met sporadically and secretly while the legal proceedings at The Hague proceeded slowly. The September 11, 2001, attacks, however, fundamentally changed the dynamics of the international security environment and ushered in with it another crisis. The United States demanded that Libya abandon its weapons of mass destruction (WMD), but, notably, President George W. Bush did not include Libya in his "Axis of Evil," nor did he reintroduce Libyan regime change as a foreign policy objective. Unlike Saddam Hussein, who blamed U.S. policies for 9/11, Qaddafi publicly condemned Al Qaeda's actions, and he agreed to recommence the stalled talks with the United States and Britain. Diplomacy made little progress, however, until September 2002, when Bush threatened Iraq over its WMD. At this point Britain proposed to Libya that, in exchange for abandoning its WMD, all sanctions would be permanently lifted and diplomatic relations restored. What differed now from the late 1990s was the credibility of the threat of military force, particularly given the success of recent U.S. operations in Afghanistan and the aggressive preparations being made to invade Iraq. These actions signaled both the capability and the resolve of the Bush administration to employ a brute force strategy to overthrow a regime. In March 2003, just days before the U.S. invasion of Iraq, Qaddafi agreed to talks over WMD. In August, Libya finally acknowledged responsibility and agreed to pay compensation for the Pan Am bombing. In return, the United States agreed to the permanent lifting of UN sanctions. The crisis ended as a coercive diplomacy success when, in October 2003, a shipment of uranium enrichment centrifuges was interdicted en route to Libya. Caught red-handed,

Qaddafi ended Libya's WMD programs while the offer of normalized relations was still on the table.

Before proceeding to examine further these three cases, a word of caution is warranted. Unlike the Iraq and Serbia crises, for which there is a variety of sources as to the perceptions and motivations of Saddam Hussein and Slobodan Milosevic, scant information is available on Qaddafi's beliefs and intentions. The following analysis therefore proceeds much more as an interpretation of how Qaddafi likely perceived the facts presented to him. This interpretation is based on his observable reactions to key events and on past research as to what motivates the charismatic leader of a personalist regime. The three conflicts presented in this chapter should therefore be considered as shadow cases.

CONFLICT WITH LIBYA

On September 1, 1969, twelve junior officers, the self-proclaimed Revolutionary Command Council (RCC), staged a bloodless coup, with Muammar Qaddafi emerging as their leader. Qaddafi's aspirations were not limited to ruling Libya alone but extended to gaining prestige and respect throughout the Arab world. After the death of Egypt's President Nasser in 1970, Qaddafi proclaimed himself successor to Pan Arabism.[2] This, in part, motivated Qaddafi to adopt his anti-Western and anti-Israeli policies, to support international terrorist groups and to seek nuclear weapons.

Qaddafi also envisioned dramatic economic and socialist reforms.[3] His most successful move came in 1973 when he nationalized foreign oil companies, the majority of which were British and American owned.[4] This, combined with the spike in global crude oil prices and an increase in Libya's oil production, generated an enormous revenue stream that allowed Qaddafi to spend liberally on modernizing Libya's military, on instituting a nuclear program, and on aid to foreign insurgents and terrorist groups.[5]

Qaddafi encouraged foreign groups in their revolutionary pursuits. By the end of the 1970s, however, his interventionist policies had alienated not only his North African neighbors but also the majority of Arab states in the Middle East.[6] In addition, Qaddafi's foreign policy was perceived by the United States as increasingly hostile to U.S. interests. Tensions escalated when the United States accused Qaddafi of supporting several international terrorist organizations, attempting to annex Chad, intervening in sub-Saharan Africa, attempting to obstruct the Middle East peace process, and trying to develop a nuclear weapons program.[7] In 1978, following Qaddafi's denunciation of the Camp David Accords, a frustrated Carter administration placed Libya on its list of state sponsors of terrorism, imposed an arms embargo, and, in February 1980, closed the U.S. embassy in Tripoli.[8]

In January 1981, President Ronald Reagan entered the White House promising to restore America's military power and prestige. Libya was low-hanging fruit for a more confrontational U.S. foreign policy, as Qaddafi was outspoken in his support of international terrorist groups. Libya had also recently invaded Chad and was actively developing nuclear and chemical weapons programs.[9] Reagan demanded that Libya change these policies though, privately, the White House did not believe such changes were likely. The administration's approach combined overt and covert military and intelligence operations, along with diplomatic and economic sanctions in an attempt to coerce, contain, and weaken Qaddafi's regime.[10]

Militarily, Reagan provoked Libya by approving naval exercises off its coast, exercises previously disapproved by Carter. These were designed to elicit a response from Qaddafi by challenging Libya's exclusive claims to the Gulf of Sidra (see Map 6.1).[11] A military confrontation ensued on August 19, 1981, when two Libyan Su-22 jets fired air-to-air missiles at U.S. Navy F-14 fighters operating south of 32° 30' latitude, Qaddafi's self-proclaimed "line of death."[12]

A month later, the Reagan administration leaked to the press that it had intelligence on a Qaddafi threat to assassinate Reagan in retaliation for the Gulf of Sidra incident.[13] The White House recalled all American citizens from Libya in December 1981 and placed a unilateral boycott on Libyan oil. Although the administration attempted to elicit international support for sanctions against Libya, their unwillingness to release any evidence of the alleged assassination plot undermined these efforts. As a result, when Reagan announced additional sanctions against Libya in February and March of 1982, even his staunchest allies refused to go along with him.[14]

Since 1972, the United States had employed a series of unilateral economic and diplomatic sanctions against Libya, all of which had reduced formal diplomatic ties and restricted trade to nominal levels. Although Libyan oil accounted for $7.8 billion, or roughly 10 percent of U.S. crude imports, and American exports to Libya were at $462 million per annum in 1980, President Reagan's sanctions forced U.S. imports to drop to a mere $9 million and exports to less than $200 million by 1985.[15]

Map 6.1. Libya.

Despite U.S. efforts to coerce Libya by military action and sanctions, over the next few years Libya continued to antagonize the Reagan administration. In April 1983, U.S. intelligence uncovered a shipment of weapons bound from Tripoli for the communist Sandinista government in Nicaragua, and, in May 1983, Libyan troops once again invaded Chad.[16] The bombing of the U.S. Marine barracks in Lebanon on October 23, 1983, though not directly linked to Libya, produced a major shift in U.S. foreign policy toward Libya.[17] In April 1984, Reagan announced National Security Decision Directive 138, establishing a more aggressive response to terrorist groups and their state sponsors.[18]

LIBYA'S SUPPORT OF INTERNATIONAL TERRORISM

Reagan's new foreign policy received an unexpected boost two weeks later when Britain severed diplomatic ties with Libya following the murder of a female police officer in front of the Libyan embassy in London.[19] Yet it would not be until the end of 1985 that the United States would consider direct military action against Libya. That year witnessed a series of hijackings and high-profile killings linked to terrorist organizations with Libyan ties, including Hezbollah, the Palestine Liberation Organization, and the Tripoli-based Abu Nidal organization. In the span of seven months, these groups carried out three airliner hijackings, one of which resulted in the murder of a U.S. sailor. They also orchestrated the seizure of the cruise ship *Achille Lauro* and the simultaneous detonation of bombs just after Christmas at the airports in Rome and Vienna, which killed twenty, including an eleven-year-old American girl.[20]

The Christmas bombings resulted in the United States escalating the conflict into a full-blown crisis. On January 7, 1986, Reagan announced "irrefutable evidence of [Qaddafi's] role in these attacks," banned all trade with Libya, and ordered all remaining U.S. nationals out of Libya.[21] In addition, Reagan ordered a second carrier group to the Mediterranean to perform operations in the Gulf of Sidra.[22] From January through March, the U.S. Navy conducted monthly "Freedom of Navigation" exercises. On March 24, Libya fired surface-to-air missiles at Navy aircraft when they crossed over the "line of death." In response, and in accordance with the White House's more aggressive rules of engagement, the Navy not only destroyed the missile site but also sank two Libyan patrol boats.[23]

Not only did Libya continue to resist U.S. demands with regard to its policies supporting international terrorism, but, after the U.S. Navy attacks, Qaddafi retaliated. On April 5 an explosion rocked a Berlin discotheque killing two, including a U.S. soldier. Most damning were encrypted messages passed between Libya's embassy in East Berlin and Tripoli prior to and following the bombing. Intercepted and deciphered by American and British intelligence, the messages directly implicated Qaddafi's regime.[24] Finally, armed with "smoking gun" evidence, Reagan ordered air strikes.

El Dorado Canyon

In the early hours of April 15, 1986, U.S. Air Force F-111s conducted a night strike, code named El Dorado Canyon, on three targets in Tripoli while U.S. Navy A-6Es simultaneously attacked two targets in Benghazi. The targeting process had commenced six months prior as intelligence and operational planners formulated military options to respond to the increase in terrorist activities linked to Libya. Following the Berlin discotheque bombing, the White House approved five targets. These were selected based on three criteria: direct links to Libyan terrorist activities, limited collateral damage potential to civilians, and minimal risk to U.S. aircrew.

Despite concerns for collateral damage, two targets were selected in urban areas. The Bad Al-Aziziyah barracks was the most lucrative target, as the compound served as headquarters for Libyan terrorist operations. It also contained Qaddafi's headquarters and residence. The other was the military barracks in Benghazi, which served as an alternative command center and provided visiting quarters for representatives of various international terrorist groups. A third target, the Murat Sidi Bilal Training Camp specialized in training commandos for attacks against naval vessels. Although it fit nicely into the targeting criteria, it had the added attraction of lying along the coast outside of Tripoli, which reduced the potential for collateral damage and limited the threat to U.S. war planes. The final two targets were airfields. One in Tripoli was home to Libya's transport aircraft, a target justified by the cargo aircraft on the ramp, which had been used to deliver weapons to support terrorist activities abroad. The other airfield, near Benghazi, was targeted to prevent MiG 23 interceptors from threatening the U.S. Navy strike package.[25]

The three Tripoli targets were allocated to the F-111s from RAF Lakenheath, England. The USAF fighter wing operated the high-tech, supersonic, terrain-following F-111Fs equipped with the Pave Tack infrared targeting system capable of delivering multiple 2,000-pound laser-guided bombs from low altitude. Prime Minister Margaret Thatcher approved the launch of the strike package from English soil. Neither France nor Spain, however, allowed the aircraft to transit their airspace, forcing them to fly a much longer, circuitous routing to Libya via the Straits of Gibraltar.[26] The extended length of the mission had an adverse impact on aircraft performance and on aircrew proficiency.

The White House billed the attacks a success, though tactically the battle damage assessment on the F-111 strikes was less than stellar. Of the eighteen fighters, one aircraft was shot down and five aircraft aborted enroute, leaving only twelve to deliver ordinance.[27] The attacks with free-fall munitions on the Tripoli airfield went relatively well, destroying two and damaging three of Libya's nine transport planes. However, of the laser-guided attacks on the naval training camp and Qaddafi's headquarters, only one-third of the bombs hit their mark. Only three of the nine jets designated for the headquarters actually employed their weapons; of these three, only two acquired their desired mean point of impact (DMPI). Although bombs fell inside the compound, none was a direct hit on the building assigned. Damage to Qaddafi's residence, however, which was not an assigned DMPI, was visible and significant. Qaddafi announced serious injuries to two of his six sons and the death of a one-and-a-half-year-old adopted daughter.[28] One aircrew misidentified its radar-offset point and released bombs onto a nearby neighborhood, killing seventeen Libyan civilians and damaging the French embassy.[29]

The U.S. Navy fared better in their twin strikes at Benghazi, with eleven of fourteen strike aircraft, all carrying free-fall ordnance, reaching their designated targets.[30] The airfield was caught off guard with its lights still on, making it relatively easy for the six A-6Es to identify and destroy six Libyan aircraft on the ramp, including two MiG 23s on ground alert. Simultaneously, five jets released their bombs on the Benghazi Military Barracks. As with Tripoli, there was also collateral damage when two bombs missed the barracks and killed five civilians in an adjacent neighborhood.[31]

Libyan Response to El Dorado Canyon

Libya immediately responded. Within twenty-four hours of the air raid, Libya launched two Scud missiles, which splashed two miles offshore a U.S. Coast Guard station in the Mediterranean 200 miles north of Tripoli.[32] Next, Libyan officials called for reprisal terrorist attacks against America and Britain.[33] Separate shootings of U.S. State Department personnel left two wounded in the Sudan and Yemen. In Beirut, one American and three British citizens were kidnapped and executed. In addition, two bomb plots were foiled in London and another two in Turkey.[34]

After the initial flurry of activity, however, the number of Libyan-supported terrorist attacks significantly decreased over time from nineteen in 1986, to seven in 1987, to only five in 1988, a reduction attributable to three factors.[35] First, European states began to enforce measures to reduce the capability of Libya to conduct further attacks. They limited the number of Libyan embassy personnel authorized to operate on their soil and restricted student visas.[36] Second, El Dorado Canyon had had a direct impact on Qaddafi, who was visibly shaken by the bombing. He withdrew from Tripoli, away from the coast, to his desert residence, where he was better insulated from potential air strikes and coup attempts. Qaddafi proved far less adversarial thereafter, publicly toning down his rhetoric. Finally, Libya's policy shifted away from planning and executing attacks and more toward covertly supporting terrorist groups.[37]

The Aftermath of El Dorado Canyon

Even though Qaddafi reduced his overt support of terrorist groups, the United States still did not achieve its core foreign policy objective of Libya abandoning its support of terrorism.[38] The United States employed a punishment strategy, using sanctions and limited air strikes to increase the costs for Libya to continue its policies. The U.S.-only sanctions proved ineffective, however, as the White House could not garner international support for a broader, multilateral approach. Reagan also made U.S. military action contingent on "smoking gun" evidence of Libyan involvement in terrorist attacks. This decision is understandable, given the diplomatic and political costs Reagan would have endured for taking further military action without such evidence, particularly because he did not have the support of the UN Security Council. The overall result of this decision, however, was a weakening in the credibility of U.S. threats skirted so long as Libya was not overtly conducting or supporting terrorist activities. Economic threats did not match U.S. demands, and, as such, no coercion range materialized wherein a negotiated settlement was preferred by Libya over continuing to resist, albeit with less anti-American rhetoric than before.

Even less successful than coercing changes in Libya's foreign policy were U.S. efforts at regime change. Despite reports of infighting within the Libyan military and several coup attempts prior to and in the immediate aftermath of the air strikes, Qaddafi remained in power.[39] Though hopeful that Qaddafi's regime would fall, by August 1986 the Reagan administration realized this was unlikely. The CIA had supported anti-Qaddafi dissident groups for over a year, but these proved disorganized, and subsequent coup attempts had been thwarted by Qaddafi's personal security force.[40] In mid-August, Reagan approved a series of disinformation operations aimed at encouraging further Libyan domestic opposition.[41] The administration discontinued its involvement in October, however, after these covert operations were made public.[42] This revelation, the easing of Qaddafi's rhetoric, the decline in the direct Libyan involvement in terrorist attacks, and the distractions to Reagan's national security team from the Iran-Contra affair stalemated the crisis until December 21, 1988, when a bomb exploded aboard Pan Am Flight 103.

The greatest hindrance to U.S. success in coercing Libya to abandon its terrorism policies was the limited level of force the United States could credibly threaten. International and domestic pressure against taking action without "smoking gun" evidence of Libyan involvement constrained the Reagan administration. The asymmetric coercion model expects that under such conditions, when a powerful challenger finds that its maximum threats are insufficient for its demands, it will either decrease its demands or abandon coercion. In this case, by allowing the crisis to stalemate, the United States, in effect, had lowered its demand from that of Libya abandoning its terrorism policies to the more modest requirement that Qaddafi not openly support terrorist activities. In the end, however, Libya would not even accede to this, as would become evident after the bombing of Pan Am Flight 103.

For this case, the survival explanation developed in Chapter 3 does not adequately explain why coercion failed. It expects coercion to fail when concession threatens the survival of the targeted state, its regime, or its regime leadership. Absent this threat, the survival hypothesis expects coercion to succeed. The following paragraphs assess the risk to the state of Libya, to Qaddafi, and to his regime for making concessions and finds that survival should not have been a significant factor in Libya's resistance.

The U.S. demand for Qaddafi to change his international terrorism policies did not threaten Libyan state survival. Compliance with this demand would have constrained foreign policy, but Libya would still have retained sovereignty over both its international and domestic policy making. Instead of threatening its survival, a concession would have actually improved Libya's external security situation. Its greatest threat came from the United States and Israel. Because the threat of military action was predicated on Libyan terrorist activity, a repudiation of such policies would have reduced the likelihood of future U.S. attacks. Israel also had the capacity to conduct limited air strikes against Libya, as it had demonstrated in 1981 with the strike on Iraq's Osirak nuclear reactor. A change in policy eliminating support for the anti-Israeli organizations of Hezbollah, the PLO, and Abu Nidal would have further decreased the likelihood of an Israeli strike. The survival of the Libyan state therefore would not have been threatened by conceding it terrorism policies but could only have been bolstered by it.

As with Libyan state survival, a concession by Qaddafi was also not likely to make his regime vulnerable to violent domestic opposition. Although a public concession would have revealed Qaddafi's regime as weakly resolved, there were no organized armed opposition groups within Libya in a position to revolt against the government. It would be over a decade before radicalized Islamic militants would begin to threaten Qaddafi's regime.

In addition to state and regime survival, a final assessment is necessary to consider whether Qaddafi's leadership position within his regime was at risk. Qaddafi ruled Libya as the dictator of a personalist regime.[43] Compared to military or single-party authoritarian regimes, or even to democratic states, leaders of personalist regimes are expected to suffer lower audience costs for making concessions.[44] This audience cost argument is derived from a principal-agent model, whereby the leader is charged with carrying out the policy preferences of powerful principals who control the regime. The principals can either reward or punish the leader by keeping him in or removing him from office. The problem for principals is that they have limited information to judge the leader's performance and must extract how well he adheres to their preferences on the basis of whether his policies succeed or fail. Audience cost is based on the leader's expectation of being removed from power for making a policy concession. For a personalist regime, audience costs are expected to be relatively low compared to other types of regimes because there are few principals likely to have the power to overthrow the leader.

According to this reasoning, Qaddafi should not have been overly concerned with the risk of a coup. On the other hand, further resistance to the United States did place his personal safety at risk, particularly if Libya continued to kill Americans. Reagan had already demonstrated a willingness to directly target Qaddafi with the El Dorado Canyon air raid. If given further evidence of Libyan terrorist actions, Reagan would likely have ordered more strikes.[45] The expectation of low audience costs, combined with the risk of further U.S. strikes for resisting, provided incentives for Qaddafi to make concessions.

Overall, the survival hypothesis incorrectly expects coercion to succeed, as concessions would not have placed the survival of Libya, Qaddafi, or his regime at risk. Libya, in fact, stood to gain security benefits by reducing the threat posed by the United States and Israel should it have abandoned support of international terrorism, as well as additional economic benefits once the Americans lifted sanctions. In addition, there were no domestic opposition groups capable of threatening Qaddafi's regime, and, as the leader of a personalist regime, his audience costs were likely to be low for making concessions. Though the survival hypothesis expected coercion to succeed, the actual outcome was a failure. What then explains Qaddafi's determination to maintain a terrorist network that would go on to take out Pan Am Flight 103 in December 1988?

The answer is twofold. The first reason has already been discussed, that is, the inability of the United States to continue to threaten Libya. The second part has to do with the weakness of Qaddafi's leadership at the time. His hold on power was more fragile than that expected for a ruler of a personalist regime. Qaddafi was clearly rattled by the bombing of his compound and retreated from public view to his desert residence. Although this action made further U.S. air strikes more difficult, it also enhanced Qaddafi's personal security. Discontent among sections of the Libyan Army and several coup attempts in the aftermath of El Dorado Canyon threatened Qaddafi's hold on power. Given this tenuous position, the additional humiliation of publicly conceding to U.S. demands might have been enough to generate internal opposition to remove him from power. Instead of conceding, Qaddafi chose the mixed response of reducing his rhetoric and limiting, but not eliminating, the number of terrorist attacks with direct Libyan involvement. He accomplished this without having to publicly concede to the United States and thus spared himself additional audience costs. Qaddafi was, in fact, concerned with his personal survival, though the survival hypothesis did not expect this to be the case.

The conventional nonrational explanations for war introduced in Chapter 2 fare somewhat better than the survival hypothesis in explaining the outcome of this crisis. Initial misperception and miscalculation by the Reagan administration over the international and domestic backlash against further air strikes explains why the White House employed a flawed coercive strategy that was, over the long run, unable to maintain credible threats to back demands. Misperception and miscalculation do not, however, explain why this crisis did not degenerate into a war. Qaddafi may have miscalculated the degree to which the administration was willing to use force prior to El Dorado Canyon. Thereafter, he rightly perceived that, so long as he toned down his rhetoric and did not provide the United States further justification for military action, Libya could continue covert actions without further escalation.

Although the rational explanation for war based on uncertainty and private information, in part, explains why the United States escalated the crisis, it fails to explain the final outcome. Uncertainty over Qaddafi's resolve to continue his terrorism policy in the face of U.S. threats was reduced over time. Following El Dorado Canyon, Qaddafi understood the conditions for which Reagan could credibly threaten force, a threshold Qaddafi carefully avoided. This decrease in uncertainty and private information did not, however, result in a negotiated settlement between the United States and Libya.

Finally, the commitment explanation expects that a weak state will resist out of concern that concessions will only lead to further demands. The United States did, in fact, have additional issues, including disputes over Libya's claims on the Gulf of Sidra, its anti-Israeli stance, and its nuclear and chemical weapons programs. The commitment explanation correctly expected Qaddafi not to publicly concede to U.S. demands. It did not, however, anticipate the partial concessions of reducing his rhetoric and limiting Libya's overt involvement in terrorist activities. According to the commitment logic, such an observable change in behavior should have been interpreted as a sign of weakening Libyan resolve and introduced an incentive for the U.S. to make additional demands, though this, in fact, never happened.

Conclusion of El Dorado Canyon Crisis

The seven-year-long conflict over Libya's policies on international terrorism ultimately ended as a coercion failure for the United States. The El Dorado Canyon air raid compelled Qaddafi to lower the rhetoric and overt support of terrorist plots to prevent further "smoking gun" evidence to justify subsequent attacks, but, overall, Libya was not convinced to abandon its terrorism policies. The Pan Am Flight 103 bombing over Lockerbie, Scotland, provides ex post evidence of the failure of the U.S. coercive strategy. Coercion failed in this case because Reagan could not continue to make credible threats of force and because Qaddafi was weak domestically. The political costs for threatening force without international support constrained the ability of the Reagan administration to use further military force. The asymmetric coercion model helps us understand what results from such a mismatch between threats and demands. The survival hypothesis fails, however, in this case as it expects coercion to succeed because Qaddafi's concessions should not have threatened the survival of Libya or his regime. Other conventional explanations for war explain why the conflict escalated to a crisis, but they fail to explain why the crisis did not further deteriorate into war. The crisis eventually deescalated into what appeared to be a stalemate until evidence linking Libya to the Pan Am bombing revealed that U.S. coercion had, in fact, failed. This revelation in 1991 led to a second confrontation, which would last another six years.

THE BOMBING OF PAN AM FLIGHT 103

By the end of President Reagan's second term in office, the crisis between the United States and Libya had stalemated. Qaddafi, who expressed a desire for more positive relations with the United States, welcomed the election of George H. W. Bush in November 1988.[46] Hopes of rapprochement evaporated, however, along with the lives of 270 passengers and crew aboard Pan Am Flight 103, which exploded over Lockerbie, Scotland, on December 21, 1988.

Accusations tying Libya to the bombing surfaced in the ensuing investigation. The break in the case came in October 1990 when the detonator for the Pan Am bomb was determined to be the same design as that employed in the later bombing of French Union des Transports Aeriens (UTA) Flight 772 over Niger on September 19, 1989. Both devices were traced to a shipment of twenty detonators purchased from Syria by Libyan intelligence. In addition, fragments of clothing in which the Pan Am bomb had been wrapped were traced to a shop in Malta. The shopkeeper identified two men as having purchased the clothing: Lamen Fhimah, station chief for Libyan Arab Airlines in Malta, and Abdel Basset, the chief of Libyan Arab Airline security. Investigators could not, however, establish a chain of responsibility for the bombing within the Libyan government and, unlike the Berlin discotheque, were never able to show that Qaddafi had prior knowledge of or had approved the attack.[47]

On completion of a nearly two-year investigation, the United States and Great Britain issued indictments on November 14, 1991, against the two Libyans, accusing them of placing the bomb in a suitcase aboard Malta Flight KM18 bound for Frankfurt, where the luggage was then transferred to Pan Am Flight 103 bound for New York via London.[48] The Bush administration adopted a coercive strategy on November 27, 1991, when they and Britain released a joint declaration demanding Libya "surrender for trial all those charged with the crime; and accept responsibility for the actions of Libyan officials; disclose all it knows of this crime, including the names of all those responsible, and allow full access to all witnesses, documents and other material evidence, including all the remaining timers; [and] pay appropriate compensation."[49]

Unlike the El Dorado Canyon crisis, however, regime change was no longer an objective. This omission makes this an interesting case for asymmetric coercion. In other cases, the United States increases its aims over time, to include homeland territory in Kosovo and to attain both regime change and homeland territory in the Iraq War. To back up these greater demands U.S. leaders ratcheted up their threats of military force. With the Libyan case, however, the United States actually reduced its objectives to no longer seek regime change. This conflict demonstrates that the United States does not always increase its demands and threats over time. In terms of the asymmetric coercion model, this reflects the idea that a powerful challenger may optimize its expected outcome by both reducing its demands and signaling costs for making threats credible.

This conflict also differs from the crises examined in this book in that it is neither a crisis nor a case of military coercion because the Bush administration did not threaten the use of force. In Chapter 2, crises were defined as conflicts between states with a heightened probability of military hostility, accompanied by a time constraint. In this particular interaction between the United States and Libya, the White House did not threaten military force, choosing instead to leverage threats of economic sanctions. This case is included, however, to assess how well the theory of asymmetric coercion developed in Chapters 2 and 3 also explains the interaction and outcome for asymmetric cases in which economic sanctions replace threats of violence. Here, rather than a punishment strategy with the threat of air strikes or a denial strategy threatening invasion, the United States employed a two-tiered punishment strategy of unilateral and multilateral sanctions. Long-term sanctions limited oil production; combined with depressed global crude prices, this resulted in a stagnated Libyan economy. The removal of these sanctions later allowed Libya to import U.S. and Western European crude oil equipment and technology to increase exports, a measure that motivated Qaddafi to eventually concede to U.S. demands.

The threat of sanctions, however, was insufficient to compel Libya to concede to all of the American and British demands laid out in the November 1991 indictment. Qaddafi declared the two suspects would not be handed over to stand trial in either America or Britain as there was no evidence of Libyan involvement in the bombing.[50] There was some indication, though, that

Qaddafi might be willing to partially concede. Libya's foreign minister denied "any Libyan connection with the aforementioned incident or any knowledge of it by the Libyan authorities." At the same time, he called "on the United States and Britain to apply the logic of law, wisdom, and reason by resorting to neutral international investigation committees or to the International Court of Justice."[51] The Bush administration quickly rejected this offer as a disingenuous stalling tactic.[52]

Libya characterized the United States and Britain as not seeking justice but rather being politically motivated because their demands would require Libya to agree to take responsibility and pay compensation prior to a verdict being reached.[53] On January 18, 1992, Libyan officials informed the UN Security Council that it was invoking Article 14 of the 1971 Montreal Convention for the Suppression of Unlawful Acts against the Safety of Civil Aviation, and Libya would conduct its own judicial proceedings in lieu of extradition.[54]

Libya's response satisfied neither the United States nor Britain, and three days later, with the support of France, they succeeded in pushing Resolution 731 through the UN Security Council. This measure denounced the Libyan government for not effectively cooperating with the investigations of Pan Am Flight 103 and UTA Flight 772. It also urged Libya to cooperate to eliminate international terrorism.[55]

In response, Qaddafi met with UN Secretary General Envoy Vasiliy Safronchuk on January 26, 1992. Following the meeting, Safronchuk announced that Libya had agreed to cooperate with the United Nations, but Libya had begun its own internal legal proceedings and would not extradite the two suspects. On February 11, the Libyan representative to the United Nations informed the Secretary General that Libya would accede to French demands regarding UTA Flight 722 and allow a French judge to travel to Tripoli to investigate the case. It would not, however, concede to American and British demands for extradition, as Qaddafi claimed this infringed on Libyan sovereignty.[56] In later talks with a UN envoy, Qaddafi cited a lack of trust as the primary reason for not allowing its citizens to be tried in either country.[57]

Despite Qaddafi's efforts to avoid sanctions, the Security Council passed Resolution 748 on March 31, 1992, which forbade the takeoff and landing of aircraft in Libya along with the supply of aircraft maintenance parts and services, prohibited the sale or transfer of arms and military equipment as well as military technical advice and training, and cut back the number of staff at Libyan diplomatic missions.[58] Notably, the sanctions did not forbid the sale of

Libyan oil, as Italy, Spain, and Germany were dependent on a continual flow of Libyan crude.[59] On November 11, 1993, with no change in Qaddafi's position, the Security Council passed Resolution 883, which froze Libyan foreign financial assets, with the exception of funds from oil sales and, more important, banned the export to Libya of selected parts and equipment to service its oil production infrastructure.[60]

Impact of Sanctions

The Libyan economy, almost entirely dependent on its oil exports, had stagnated in the late 1980s as the global price of crude fell precipitously. Oil prices would remain depressed from 1986 until 2003, while Libya's oil production remained relatively fixed at 1.5 million barrels per day from 1991 to 2003, even after OPEC raised Libya's quota.[61] As a result, Libya's per capita gross domestic product (GDP) slowly eroded from $11,200 in 1991 to $9,200 by 2002.

The major economic impact of sanctions fell on Libya's ability to import goods, a strain that placed inflationary pressure on Libya's markets.[62] In the long run, the sanctions made it difficult for the oil industry to procure the equipment and spare parts to maintain its infrastructure and made it impossible to increase production. This inflexibility in adjusting oil exports to offset low oil prices left Libya's economy even more susceptible to price fluctuations in the global oil market. Even so, Libya was fortunate relative to Iraq because the sanctions did not boycott oil exports altogether, thus sparing Libya the economic meltdown suffered by Iraq.

The exclusion of oil exports from UN sanctions is an important aspect to this case. Unlike Iraq's economy, which had by the mid-1990s collapsed under the weight of the oil embargo, Libya's economy was spared such a fate. Its economy instead suffered a gradual decline. Unlike Saddam Hussein, who by 1994 had little to lose in confronting the United States, Qaddafi still had much of the Libyan economy he could salvage by acceding to demands, and because regime change was not at stake he could do so and remain in power.

An indirect impact of the sanctions, however, was the threat posed to Qaddafi's regime by the rise of radicalized Islamist groups within Libya. Unemployment hovering at 20 percent, double-digit inflation, and a youth bulge with 70 percent of the population under the age of twenty all contributed to a growing disillusionment with the regime. This, in turn, prompted some of Libya's youth to join the militant opposition groups spreading throughout Arab states, including organizations with ties to Al Qaeda.[63]

In June 1995, violence erupted in the form of attacks on government security forces in Benghazi and the central region of Libya. In August, an assassination attempt on Qaddafi failed, but the sporadic violence continued, including another attempt on his life in 1996 near his hometown of Sirte.[64] In 1998, Qaddafi responded to the insurgency, sending a thousand troops into Benghazi to crush the uprising.[65] The threat to his regime from these radicalized groups appears to have had a sobering affect on Qaddafi, who now reversed his previous policy of supporting terrorism and took action against terrorist groups within Libya. As a result, by the late 1990s Qaddafi's regime was much less vulnerable to internal opposition. Despite this shift away from supporting international terrorism, Qaddafi did not change his position on extraditing the two Pan Am Flight 103 suspects.

U.S. Response and Qaddafi's Diplomatic Triumphs

Back in 1995, it had become apparent to the United States that the sanctions on Libya were insufficient to sway Qaddafi. To increase pressure, Congress extended the provisions of the "Iran Foreign Oil Sanctions Act of 1995" to also include Libya. This bill imposed sanctions on foreign companies that invested more than $40 million per annum in Libya's petroleum industries.[66] Another congressional action was to pass the "Antiterrorism and Effective Death Penalty Act of 1996." This law modified federal statutes, allowing the relatives of the 105 U.S. victims of Pan Am Flight 103 to bring a $4 billion civil suit against the Libyan government.[67] This legislation would eventually provide the mechanism for Libya to compensate the families of the victims.

Though UN sanctions did not debilitate Libya's economy as it had in Iraq, nearly five years of economic sanctions had weakened Libya and isolated it from the world community. Diplomatically, however, the tide finally began to turn in its favor in 1997 and 1998 as Qaddafi achieved a series of diplomatic triumphs. In August 1997, he traveled to Niger and met with the presidents of Burkina Faso, Chad, Mali, and Niger. These regional leaders issued a joint statement announcing closer economic cooperation among their countries and calling on the United Nations to evaluate the impact of sanctions on Libya.[68] Qaddafi made further progress in September when the Arab League passed a resolution allowing both diplomatic and humanitarian flights to and from Libya and releasing Libyan funds held in Arab banks.[69] One of the most significant diplomatic gains, however, came in February 1998 when the International Court of Justice (ICJ) ruled against the United States and Britain and declared that it alone had the jurisdiction to decide whether Libya must surrender the two accused.[70]

Two factors account for Libya's diplomatic success. First, by the late 1990s Qaddafi had reversed his most aggressive foreign policies by no longer intervening in neighboring countries and by withdrawing his support for international terrorist groups. Libya was now less of a threat to African and the Middle Eastern countries. Second, there was a broader international backlash against the use of economic sanctions. This followed, in part, from UNICEF reports in the mid-1990s of the steep rise in Iraq's infant mortality rates. As a result, many countries were now less willing to enforce sanctions against Libya.

Nearly seven years after rejecting Libya's compromise proposal for a trial by the ICJ, the United States and Britain reversed their position. On August 24, 1998, they issued a joint letter to the UN Security Council with an initiative to try the two Libyan suspects at The Hague. They further agreed that, on Libya's delivery of the two accused, UN sanctions would be suspended, though not yet permanently lifted.[71]

Four reasons explain this reversal in policy. First, it had long been clear that sanctions alone would not compel Qaddafi to extradite the Libyan suspects to America or Britain. In addition, U.S. efforts to strengthen economic sanctions on Libyan oil exports had failed, and future attempts were also not likely to succeed. Second, current sanctions on Libya were unraveling in the face of international reactions against Iraqi sanctions, with African and Arab countries now openly defying the existing Security Council resolutions. Third, the British Labour Party held the majority in Parliament with Tony Blair as their new prime minister. Unlike John Major, Blair had not been personally involved with the tragic events of the Pan Am bombing, and he proved more willing to consider a compromise. Finally, the British families of the victims, some of whom had met personally with Qaddafi, supported the third-country legal framework. In the summer of 1998 they placed pressure on Blair, who, in turn, convinced Clinton to support a trial at The Hague.[72]

Conclusion of the Pan Am Flight 103 Case

Though all parties were now in agreement in principle on extradition, there were no formal diplomatic ties between Libya and the United States or Britain to commence negotiations. After mediation by UN Secretary-General Kofi Annan and South African President Nelson Mandela, however, Qaddafi finally handed the two suspects over on April 5, 1999.[73] As promised, UN sanctions were then suspended, though not formally lifted until August 2003. The United States maintained its unilateral sanctions until Libya would agree to take responsibility and pay compensation for the bombing and also added demands that Qaddafi publicly renounce support of international terrorism and abandon Libya's WMD programs.[74]

This case ends in April 1999 when Libya conceded to the U.S. core demand for the extradition of the two Libyan suspects. Qaddafi did not, however, agree to the entirety of the demands, as he still refused to send the men to America or Britain. In addition, though Libya would go on to cooperate with the investigation, Qaddafi refused to take responsibility for the bombing and to pay compensation prior to a verdict being reached. Economic coercion therefore proved to be successful, but only partially. Although the crisis deescalated with the extradition and subsequent suspension of UN sanctions, the conflict would continue until responsibility and compensation for the bombing could be settled and two other issues, Libya's terrorism policies and its weapons of mass destruction (WMD) programs, could be resolved.

Explaining Economic Coercion Success

Under the equilibrium conditions for the asymmetric coercion model developed in Chapter 2, coercion should always succeed. Rationally, a powerful challenger should choose a coercive strategy only when it expects a weak target will concede and then only when the expected outcome of coercion is better than the alternatives of accommodation or brute force. The coercion range identifies the set of successful demands for which both the challenger prefers coercion to brute force and the target prefers concession to resistance. Instead of military force, however, in the case of Pan Am Flight 103, the United States employed economic sanctions. What can the asymmetric coercion model tell us about cases in which there is no threat of violence?

As this case suggests, the logic of coercion remains relevant when sanctions replace the threat of force. When violence is not a viable option, then a brute force strategy is not available, which leaves the choice between economic coercion and accommodation. Again, a powerful challenger should choose coercion only when it is preferred to accommodation, and this is the situation only when the target is expected to concede. If the target were to resist instead, the challenger would gain none of its objectives despite having incurred the sunk costs for signaling the credibility of its economic threats.

The survival hypothesis anticipates that the equilibrium conditions of the asymmetric coercion model will still hold. Coercion should succeed unless concession threatens the survival of the targeted state, the ruling regime, or its leader. With regard to Libya, compliance with demands to extradite, co-operate, acknowledge responsibility, and compensate for the bombing did not threaten state survival. State survival, as explained in Chapter 3, requires that a state maintain its international and domestic sovereignty, that is, con-trol over its foreign policy making and its homeland territory, respectively. Conceding to U.S. demands would not have weakened Libya's control over its foreign policy making. Domestically, although Qaddafi rightly viewed the demands as an infringement on Libya's sovereignty, the concessions had no impact on its control over its homeland territory. As a result, Libya could con-cede to U.S. demands without a threat to the state. In fact, a concession could have improved the economic security of Libya by securing a permanent lifting of sanctions.

In terms of regime survival, a concession by Qaddafi in 1991 would not have placed his regime at risk from domestic opposition. Although a pub-lic concession would have revealed his regime as weak, there were then no militant groups within Libya poised to revolt. By 1995, however, a weakened Libyan economy had generated a number of unemployed and discontented youth susceptible to radicalization. Concern over looking weak to militant groups may, in part, explain why Qaddafi would make no further compro-mise, even though Libya suffered economically. The window of opportunity for a rebellion was short-lived, however, and by 1998 loyal troops had crushed the opposition and removed the domestic threat to Qaddafi's regime.

Along with state and regime survival, an assessment is also required as to whether concessions would have risked Qaddafi's leadership within his regime. As ruler of a personalist regime, he should have been less likely to incur significant audience costs for conceding and admitting to failure. As demonstrated during the 1980s, however, Qaddafi had proven vulnerable to coup attempts arising from internal machinations. Aside from the fact that he continued to rule, there is scant evidence to determine the degree to which Qaddafi was concerned about audience costs, although he did take two mea-sures to avoid such costs. First was the manner in which Libya rejected U.S. and British demands. Although Qaddafi publicly rejected the joint proposal in November 1991, Libya's foreign minister set conditions for extradition to a third country for trial. Qaddafi had not rejected the demand of extradition, only where the trial would take place. By stipulating these conditions from the outset, it would be the United States and not Libya that would later be seen as making concessions, despite the fact that the United States still obtained its core objective.

The second way Qaddafi offset audience costs was by deferring extradition until his power and prestige was on the rise. He agreed to demands in 1999, once the militarized Islamic opposition had been crushed and his regime was again firmly in control of the country. His recent diplomatic victories in Africa and the Middle East had also bolstered his stature, making it easier for him to absorb criticism from within the regime for allowing the two officials to be extradited.

In conclusion, neither the survival of the Libyan state nor that of Qaddafi's regime was at stake by conceding to demands. There is insufficient evidence, however, to conclude as to whether Qaddafi's personal survival was threatened. His refusal to extradite the two suspects to America or Britain, or to acknowledge responsibility for the bombing and pay compensation, indicate that audience costs for making such concessions may have been a factor. It may therefore have taken the explicit agreement of the United States to a trial at The Hague and its implicit acceptance of Libya's refusal to take responsibility for the bombing or pay compensation until after the trial before Qaddafi felt comfortable enough to concede.

While the survival hypothesis correctly expected coercion success, the conventional explanations for bargaining failure and war introduced in Chapter 2: misperception and miscalculation, uncertainty and private information, and commitment problems do not help us understand the outcome of this crisis. As with the El Dorado Canyon crisis, miscalculations arising from misperception and asymmetric information better explain the initiation and length of the crisis rather than the final outcome. The United States and Britain miscalculated the impact on Qaddafi of the extradition of Libyan suspects to their soil and the act of taking responsibility and paying compensation prior to a verdict. Qaddafi, likewise, miscalculated the long-term impact of U.S. sanctions on Libyan oil production, and no one could have forecast the depressed global oil prices, which lasted throughout this lengthy case. Finally, the commitment explanation for war expected Libya to resist extradition out of concern that it would only lead to further demands over the issues of Libya's terrorism policies and its WMD programs. Qaddafi, however, did concede to extradition despite knowing the United States would introduce additional issues that would have to be dealt with before U.S. sanctions could be removed.

U.S. and Libyan Relations from 1999 to 2001

Tensions between the United States and Libya abated with the extradition of the two suspects of the Pan Am Flight 103 bombing. For its part, Britain reinstated diplomatic ties with Libya in July 1999, but the United States would not consider such a move nor lift its unilateral sanctions until additional concerns were addressed.[75] In regard to Pan Am Flight 103, Qaddafi had yet to accept responsibility for the bombing or pay compensation. In addition, there was the lingering issue of Qaddafi publicly renouncing his support of terrorism. Finally, the White House raised concerns over Libya's WMD as evidence began to surface during the late 1990s that Libya was attempting to reenergize its nuclear program.[76]

Though diplomatically Libya had shed much of its pariah status to the rest of the world, this did not translate into a significant increase in trade or growth in Libya's long-dormant economy. With depressed global oil prices, economic recovery was dependent on a boost in oil production, an option limited by Libya's undercapitalized oil industry, which had fallen into neglect under years of sanctions. Its oil production and transportation network had originally been constructed by American companies in the 1960s and 1970s. An increase in the flow of crude now required the injection of U.S. equipment and technology. Though Qaddafi had succeeded in having UN sanctions suspended, the resolution had not been rescinded. Uncertainty over whether the UN Security Council might reinstate sanctions, along with U.S. sanctions still in place, kept away leery Western investors. With the objective of having sanctions removed, Libya commenced negotiations with the United States in May 1999.

The Clinton administration agreed to talks contingent on their being held in secret.[77] Though U.S. objectives remained classified, a November 1999 speech by former U.S. Deputy Assistant Secretary of State Ronald Neumann revealed the United States' overall foreign policy goals with regards to Libya. Prior to the removal of unilateral sanctions, the administration expected a final settlement of the Pan Am bombing, that Qaddafi take steps to publicly disavow himself of terrorism, and that Libya cease the pursuit of WMD and missile programs. Neumann maintained that Libyan regime change was not a U.S. objective.[78] The private talks had made some progress when the White

House suspended them in early 2000 over concerns that knowledge of the talks would be leaked and disrupt the upcoming presidential election.[79]

Although the previous two cases discussed Libya's involvement with international terrorism, little has yet been mentioned concerning Libya's nuclear, biological, and chemical programs. The next section reviews the history and motivations for Libya's development of these unconventional weapons programs.

QADDAFI'S NUCLEAR, BIOLOGICAL, AND CHEMICAL WEAPONS AMBITIONS

Libya's quest for WMD began with Qaddafi's rise to power in 1969, and his efforts continued for more than three decades. From the outset, Qaddafi placed high priority on obtaining nuclear weapons. Given the great wealth generated from the influx of oil revenues in the 1970s, Libya's initial efforts were directed at purchasing weapons from nuclear states, thereby circumventing the arduous process of developing an indigenous nuclear program. Libya first approached China in 1969 and, in the late 1970s, made overtures to the Soviet Union, France, and India.[80] When these efforts failed, Qaddafi instituted Libya's Nuclear Energy Commission in 1973 with the mandate of developing the scientific and technical capacity for a domestic nuclear program.[81] Libya took a step forward in 1975 with the purchase of a nuclear research reactor from the Soviet Union, which achieved operational status in 1981.[82]

The initial motivation for Libya to acquire nuclear status was based on Qaddafi's desire to gain prestige and respect within the Arab world. After the death of Egyptian President Nasser in 1970, Qaddafi proclaimed himself successor to Pan Arabism, much to the dismay and derision of leaders of other Arab states.[83] Libya's nuclear weapons program, along with its ongoing procurement of modern conventional weapons, which included Soviet tanks, jets, and air defenses, was, in part, an effort to bolster Qaddafi's image in North Africa and the Middle East.

A second motivation for acquiring nuclear weapons lay in Libya's growing security concerns over Israeli conventional air power and nuclear weapons capabilities. Relations with Israel deteriorated rapidly after Qaddafi's rise to power and remained strained throughout the 1970s.

By the 1980s, Libya's nuclear program had stagnated in light of concerns over Qaddafi's intentions to weaponize. Unable to form a cadre of Libyan scientists and other skilled technicians, Qaddafi remained dependent on foreign expertise.[84] By the mid-1980s the Soviets were concerned with Libya's desire to enrich uranium to weapons-grade levels and, as a result, were no longer willing to offer scientific and technical advice or sell equipment and parts. Other countries were likewise discouraged from providing Libya with support. Qaddafi, for his part, made no effort to hide his desire for nuclear weapons and long-range missiles.[85]

In contrast to its stalled nuclear program, Libya's chemical weapons program advanced rapidly during the 1980s, due in large part to the technical advice of West German companies. Libya constructed three chemical factories during the 1980s, its first and best known being the Rabta facility seventy-five miles south of Tripoli.[86] This plant produced blister and nerve agents that were employed against Chad troops in 1987.[87]

In spite of its active chemical weapons program, Libya publicly disavowed biological weapons, however, and signed the international Biological and Toxic Weapons Convention (BTWC) in 1982. Although accusations surfaced in the 1990s that Libya had attempted to acquire biological weapons, no evidence of such a program has been uncovered.[88]

In the late 1990s, Libya's dormant nuclear program was reinvigorated with the influx of weapon designs and centrifuge equipment, courtesy of the A. Q. Khan network.[89] Libya spent as much as half a billion dollars over six years to acquire the technical knowledge and equipment to design and build a nuclear device.[90] But, despite this direct injection of technology, a combination of poor management practices and the lack of domestic scientific and technical expertise plagued Libya's ambitions. It failed to mature the key capabilities of uranium enrichment, weapons design, and long-range missile production.[91]

The El Dorado Canyon air raid in 1986 demonstrated to Qaddafi that the United States had the capability to also strike his WMD facilities in the same manner that Israel had destroyed Iraq's Osirak nuclear reactor in 1981. As a result, Libya constructed its next two chemical facilities underground in hardened tunnels.[92] In 1990, the Libyans set fire to tires set on top of the exposed Rabta plant to trick U.S. intelligence into assessing the plant as nonoperational.[93]

With regards to biological weapons, though Libya had signed the BTWC, this did not assuage U.S. concerns. In the mid-1990s, the Clinton administration alleged that Libya had made efforts to acquire biological weapons technology from South Africa. At the same time, the administration conceded that, even if Libya had received help, its scientific and technological constraints seriously hindered the future development of a biological weapons program or weapons delivery system.[94] The United States assessed that, although Qaddafi may have wanted biological weapons, Libya neither had such a program nor possessed the know-how to develop one.

CRISIS OVER LIBYA'S WEAPONS OF MASS DESTRUCTION

The newly elected President George W. Bush was surprised to learn of the secret talks with Libya and was reluctant to reinitiate talks even after The Hague reached a verdict on January 31, 2001, convicting one of the two Libyan officials and acquitting the other. The Bush administration feared the political fallout should it have become known that they were secretly negotiating with Libya.[95]

The conflict between the United States and Libya changed dramatically, however, following the September 11, 2001, attacks, when Bush escalated U.S. security concerns over WMD getting into the hands of international terrorists. This generated a crisis between the United States and Libya. The Bush administration's demands remained unchanged from those articulated by the Clinton administration in 1999: to acknowledge responsibility and pay compensation for the Lockerbie bombing, to renounce Libya's support of terrorism, and to abandon its WMD programs. It was the last of these demands that comprised the core U.S. objective and served as the catalyst for this new crisis.

The fundamental change in the international security environment following the 9/11 attacks was not lost on Qaddafi, who immediately undertook renewed efforts at rapprochement. He had already altered his policy toward terrorism in the late 1990s following threats to his regime from radicalized Islamic groups. Libya's terrorism policies were therefore now aligned with those of the United States. These factors made Libya one of the first and most vocal of Arab states to condemn Al Qaeda and offer intelligence to assist the United States.[96]

In October 2001, the White House recommended negotiations with Libya, though these talks moved slowly as the administration focused on Afghanistan.[97] In May 2002, Libya offered to pay $2.7 billion to the families of Pan Am bombing victims, with a partial payment to be made after the permanent lifting of UN sanctions, a second payment contingent on the suspension of unilateral U.S. sanctions, and a final payment to be made once the U.S. State Department removed Libya from its list of state sponsors of terrorism.[98] The

United States rejected this offer, however, as it failed to address Qaddafi taking responsibility for the bombing and Libya's WMD programs.

In September 2002, Bush spoke before the UN General Assembly where he demanded Iraq abandon WMD or face regime change. Although the United States' attention was on Iraq, British Prime Minister Tony Blair, with Bush's blessing, approached Qaddafi with an offer to normalize relations between the United States and Libya if Qaddafi abandoned its WMD.[99] By this time, Libya had clearly reversed its policies such that terrorism was no longer an issue.

Although the demands communicated by Blair remained largely unchanged from earlier U.S. offers, the threat level to back these demands had increased markedly. The threat of economic sanctions had been lowered substantially since UN sanctions had been suspended in 1999, and only unilateral U.S. sanctions remained in place. The overall threat level was boosted dramatically, however, by the increased credibility of the United States taking military action. Though Bush did not directly threaten Libya with either air strikes or invasion, he did not have to, as the successful battles waged in Afghanistan in 2001 demonstrated to Qaddafi the ease with which the United States could employ its forces against Libya should it not abandon its WMD.

For six months after Blair's overture, however, Libya remained silent until the eve of the U.S. invasion of Iraq in March 2003. Libyan officials then contacted Britain, agreeing to abandon WMD in exchange for the removal of sanctions and the reestablishment of diplomatic and economic ties with the United States.[100] Tripartite negotiations among the United States, Britain, and Libya recommenced in late spring of 2003 with the objective of reaching a final resolution to the Pan Am Flight 103 bombing and Libya's WMD.[101] In August, a Libyan official delivered a letter to the UN Security Council, agreeing to take responsibility for the Pan Am bombing and to pay compensation.[102] In return, the United States announced it would not oppose ending UN sanctions against Libya and, on September 12, 2003, the Security Council adopted UNSCR 1506, permanently lifting the 1992 sanctions.[103] Still, the Bush administration kept unilateral sanctions in place until a final resolution could be reached over Libya's WMD.

Libya's concessions came in a series of tit-for-tat exchanges. In the immediate aftermath of the September 11, 2001, attacks, Qaddafi publicly condemned Al Qaeda and provided intelligence to the United States. Such cooperation, which Libya had demonstrated even before September 2001, continued until, by the fall of 2002, it was clear to the United States that Qaddafi no longer supported international terrorism. This and pressure by Blair convinced Bush to exclude Libya from his "Axis of Evil" list of rogue states and, unlike the case of Iraq, not to seek regime change. The second stage of concessions came in August 2003 when the remaining issues from the Pan Am bombing were finally resolved: The Libyan government issued a statement taking responsibility and agreeing to pay the $2.7 billion settlement to the families of the victims: in response, the UN Security Council permanently removed sanctions against Libya.

The final concession by Qaddafi, that of abandoning Libya's WMD, was the result of British and U.S. intelligence sources identifying, tracking, and intercepting an A. Q. Khan shipment of uranium enrichment centrifuges aboard a ship bound for Libya in October 2003. This action accelerated the trilateral negotiations with Libya, providing U.S. and British intelligence agencies access to Libyan chemical and nuclear sites. In mid-December Libya offered to renounce its WMD programs, and on December 19 Qaddafi made this decision public.[104] The crisis concluded in the ensuing days when Libya, as promised, opened its chemical and nuclear facilities to IAEA inspectors. By June 2004 the United States, also as promised, had removed sanctions and resumed diplomatic relations with Libya.

Why Libya Gave up Its WMD

Libya eventually met all U.S. conditions to bring this crisis to a conclusion. In terms of the asymmetric coercion model, the United States maintained the same level of demands, but after the September 11, 2001, attacks the credibility of its threats was dramatically increased from that of unilateral economic sanctions to that of a conventional military attack. This raising of threat level to meet demands convinced Libya to concede and coercion to succeed. What remains contentious is precisely why Qaddafi chose to abandon his long-standing nuclear ambitions. Three arguments have emerged to explain why he chose to concede when he did. George W. Bush and Dick Cheney articulated one argument in their bid for reelection in 2004 by taking credit for their administration's preventive war doctrine and the impact the 2003 invasion of Iraq had on Qaddafi.[105] A second argument by the Clinton administration contends that the diplomatic channels forged during the late 1990s set the stage for Qaddafi to concede WMD.[106] A third argument focuses on the change in Libyan interests, brought on by sanctions, a weak economy, and growing dissent within the state that translated into a desire of Qaddafi to shed Libya's pariah status and normalize relations with the United States.[107] Individually, each of these arguments identifies necessary, but not sufficient, reasons for Qaddafi to abandon his WMD programs. The successful outcome appears to have resulted from all three factors along with Libya's inability to develop an indigenous nuclear program. Further, the key decisions by the United States not to seek regime change and to negotiate a settlement whereby concessions could be made incrementally in tit-for-tat exchanges made the process more palatable for Qaddafi to come to terms.[108]

The argument that Bush's aggressive policies following the September 11 attacks caused Qaddafi to concede his WMD rests heavily on the timing of key Libyan decisions. Following 9/11, talks between the United States and Libya recommenced after a hiatus of a year and a half. Libya, however, was unwilling to link the issue of WMD to a final settlement of the Pan Am Flight 103 bombing, and talks stalled until mid-March 2003, as the United States made its final preparations to invade Iraq. The timing suggests Qaddafi agreed to abandon WMD out of fear that a similar fate would befall Libya.[109] In addition, as Cheney would later underscore, Qaddafi's December 19, 2003, announcement to abandon WMD came just four days after the U.S. military had captured Saddam Hussein.

Qaddafi's decision to enter trilateral negotiations over WMD in March 2003 coincided with the United States demonstrating both its military capability and its resolve to punish and overthrow regimes that supported terrorism or allegedly possessed WMD. The U.S. military toppled the Taliban in Afghanistan in 2001 and had deployed a hundred thousand troops to the Middle East in preparation for an invasion of Iraq. The threat of U.S. military action against Libya was made credible by the success of operations in Afghanistan and by the temerity of the United States, which now appeared willing to invade Iraq even without authorization from the UN Security Council. Given this demonstration of capability and resolve, it is reasonable to conclude that U.S. actions against Afghanistan and Iraq motivated Qaddafi in his decision to reinitiate negotiations with the U.S. and Britain.

It is, however, important to understand the terms for these renewed negotiations. Six months earlier, in September 2002, Prime Minister Blair had extended an offer to Qaddafi to normalize relations in exchange for Libya ending its WMD program. Accepting this offer could have solved two problems for Libya. First, the lifting of economic sanctions and reestablishment of diplomatic relations with the United States could improve Libya's economy. This had for over a decade been a foreign policy objective for Qaddafi. Second, the abandonment of WMD would remove a casus belli for potential U.S. military action against Libya. Forsaking WMD would cost Libya its current chemical weapons program and the hope that some day it could develop nuclear weapons. Unlike the promise of nuclear weapons, however, chemical weapons had only a limited deterrent capability. Given the limited benefits Libya expected from keeping its WMD program, accepting the terms then would have improved both Libya's security situation and its economic outlook.

The second point made by Cheney, that is, that Qaddafi's decision to abandon WMD was the result of the capture of Saddam Hussein, is, however, unfounded. By mid-December 2003, prior to Saddam's capture, Libya had already agreed to abandon its WMD in exchange for normalized U.S. relations, though the public announcement by Qaddafi did not take place until December 19. There is no evidence that the capture of Saddam Hussein played a role in Qaddafi's decision making. Those who discount the impact of Bush's preventive doctrine on Libya's decision to abandon its WMD argue instead that Libya had been willing to concede its WMD well before 2003 but that it was U.S. policies that had delayed an earlier settlement.

A second argument maintains that the diplomatic channels and trust forged during secret negotiations of the late 1990s provided an earlier opportunity to resolve the issues between the United States and Libya. It was the delay in negotiations caused by the Americans that prevented a resolution that could have headed off this crisis. Qaddafi had indicated a willingness to concede his WMD programs as early as 1999. According to Martin Indyk, who served as a senior director on Clinton's National Security Council and participated in the secret talks, Libya was willing to discuss any issue with the United States.[110] The Clinton administration, however, set the agenda to first settle the Pan Am affair before broaching the broader issue of Libyan support of terrorism and its WMD programs. According to Indyk, this process was later derailed by the Bush administration's refusal to reconvene talks.[111]

Qaddafi clearly was eager to have the remaining sanctions removed, but he had not indicated a willingness to concede to all U.S. demands. In 1999 Qaddafi was unwilling to consider paying any compensation until after a verdict in the Pan Am trial had been rendered. Also, Libya refused to negotiate directly over WMD, offering instead to participate in a multilateral forum. In May 2002, Libya finally proposed to pay compensation in exchange for the lifting of sanctions and the removal of Libya from the U.S. list of state sponsors of terrorism. The proposal made no mention of WMD, however, and thereafter Libya never indicated a willingness to directly negotiate with the United States over WMD until March 2003.

Though Libya was willing to continue negotiations, it was the United States that stagnated the process. The Clinton administration established the sequential strategy, resulting in a delay in talks on WMD. In 1999, following the extradition of the two Libyan suspects, Clinton agreed to the suspension of UN sanctions. The White House then planned to permanently lift the UN sanctions in exchange for a final resolution on the Pan Am bombing. Once the Pan Am case had been settled, the United States would then consider lifting its unilateral sanctions and normalizing relations in exchange for Libya abandoning WMD and its support of terrorism. The subsequent suspension of talks by the White House during the 2000 presidential election and the delay by the Bush administration in reinitiating talks until after the September 11 attacks only frustrated the process. Further negotiations then floundered because of Qaddafi's reluctance to take responsibility and pay compensation for the Pan Am bombing or to begin direct negotiations with the United States over WMD.

Although the diplomatic progress in the late 1990s did establish negotiating channels and a modicum of trust between the United States and Libya for further talks, it was Qaddafi who was unwilling to negotiate over the remaining demands until the eve of the U.S. invasion of Iraq. The diplomatic channels forged through British efforts therefore proved necessary but insufficient to prevent this crisis.

A third argument points out that Libya had long wanted to end its conflict with the United States. Qaddafi had desired rapprochement since Reagan departed office, but the diplomatic fallout from Pan Am Flight 103 proved a major stumbling block to this effort.[112] In addition, the impact of sanctions on Libya's stagnating economy during the 1990s and the rise of radicalized Islamic terrorist groups further convinced Qaddafi that it was in his self-interest to resolve the conflict.

A shortcoming of this view that Qaddafi had changed his anti-Western stance by the late 1980s is that it ignores the fact that the Pan Am bombing was a byproduct of Libya's terrorism policies. It was not until 1991, when confronted with evidence that Libyan officials were directly involved in the bombing, that Qaddafi began to take steps to reverse these policies. Despite the pressure from the United States, it was the radicalization of Islamic groups against his regime in the late 1990s that most likely contributed to the reversal of his terrorism policies.

A second weakness in this line of reasoning is that, although Qaddafi may have desired rapprochement, he was unwilling to meet demands to extradite the two suspects to America or Britain. It would be the United States and Britain who eventually compromised and agreed to Libya's offer for a trial at The Hague. If rapprochement had been truly important to Qaddafi, he could have conceded at any time in the seven years since the United States and Britain introduced its 1991 joint indictment.

A final critique of this argument questions why, if Qaddafi's interests in WMD had changed, did Libya secretly procure upwards of a half a billion dollars worth of nuclear technology from the A. Q. Khan network after 1997? One speculation is that Qaddafi continued with his nuclear program to provide a bargaining chip for future negotiations.[113] This seems unlikely. To be used as leverage would have required Libya to eventually make known its nuclear acquisitions, something Libya never made any effort to do. More puzzling is the fact that Libya continued to make purchases from the A. Q. Khan network even after the United States invaded Iraq. Qaddafi likely realized that the revelation of such transactions would, if anything, reduce Libya's bargaining leverage rather than enhance it. Indeed, that is precisely what happened on its discovery in October 2003.

The three previous arguments provide insightful but insufficient reasoning for why a final agreement was reached in December 2003.[114] By the end of the 1990s, sanctions and the rise of militant Islamic groups within Libya had likely persuaded Qaddafi to abandon his support of terrorism. This had not, however, been sufficient to cause him to extradite the two Pan Am suspects to America or Britain. Though the White House finally agreed to a trial at The Hague, further negotiations failed to convince Qaddafi to acknowledge responsibility and pay compensation for the Pan Am bombing or agree to negotiate over WMD. Finally, the U.S. invasion of Iraq likely motivated Qaddafi to negotiate over the remaining U.S. demands, but it was the intercept of uranium enrichment centrifuges that finally convinced him to abandon his WMD. At this point, Qaddafi took the deal while it was still available, exchanging WMD, which consisted only of chemical weapons at this point, for normalized relations with the United States.

A final point not discussed in the three previous arguments is the importance of the U.S. decision not to make regime change a foreign policy objective.

The willingness of Bush to agree to Blair's request to keep the Undersecretary of State for Arms Control, John Bolton, out of negotiations relieved pressure from neoconservatives within the administration to increase demands on Libya.[115] The further willingness of the United States to coordinate and abide by the timing and wording of public announcements concerning Libyan concessions provided Qaddafi a means to save face and thereby reduced his audience costs.

Survival Concerns of Libya and Qaddafi

The previous section argues that a combination of factors contributed to U.S. coercion success. In terms of the asymmetric coercion model, after the Al Qaeda attacks on September 11, the United States would no longer accommodate Libya's WMD. Bush escalated the conflict into a crisis, but, given the priority of military operations first in Afghanistan and then in Iraq, he chose a coercive strategy for Libya over the alternative of brute force. Even with the escalation of the conflict, the United States did not increase its demands but maintained settling the Pan Am case and dismantling Libya's WMD programs as its core goals. It is important that the Bush administration did not add regime change as a foreign policy objective, which would have directly threatened Qaddafi's regime survival. To back up its limited demands the United States significantly increased the threat of military force. Operations in Afghanistan and the deployment of forces to the Middle East in preparation for the invasion of Iraq were costly signals sent to Qaddafi, making credible the threat of the United States taking decisive military action. A settlement between the United States and Libya could now take place; that is, a coercion range had materialized wherein the United States preferred a negotiated outcome to a brute force invasion, and Libya preferred concession to resistance and the risk of U.S. military action. The survival hypothesis expects that, under the equilibrium conditions of the asymmetric coercion model, a targeted state will acquiesce and coercion will succeed, as long as concession does not place the survival of the state, its regime, or its regime's leadership at risk.

A state's survival depends on it maintaining domestic and international sovereignty, measured by its control over homeland territory and foreign policy decision making, respectively. Settling the Pan Am bombing and the issue of Libya's WMD threatened neither. Conceding had no impact on Libya's ability to control its homeland territory as it only had chemical weapons with minimal operational capability and little deterrent value, as discussed in the following paragraphs. Although abandoning WMD did constrain Libya's foreign policy, it was Qaddafi who made that choice. Agreeing to U.S. demands therefore did not threaten the survival of Libya but would, in fact, increase its security and economic outlook through normalized relations with the United States.

Abandoning its WMD programs therefore had little negative impact on Libya's security. With regard to nuclear weapons, Qaddafi had originally sought a nuclear program for prestige and security. Libya would have become the only Arab country with the same nuclear capabilities as Israel. After the El Dorado Canyon air raid, Qaddafi presumed that nuclear weapons would also provide a deterrent against further U.S. strikes. This calculus reversed itself abruptly, however, on September 11, 2001. Rather than dissuading, Libya's nuclear program now encouraged U.S. aggression, as had Iraq's alleged WMD. Because Libya had been unable to develop an indigenous nuclear weapons program and future development seemed unlikely, abandoning a nonfunctional nuclear program did not forfeit security but actually increased it by removing a casus belli for a U.S. attack. Likewise, chemical weapons had little, if any, deterrent value against a U.S. military well prepared to fight in a chemical environment, as demonstrated in the 1991 Gulf War.

In addition to the survival of Libya, the survival hypothesis expected Qaddafi to be concerned about the survival of his regime in the face of violent domestic opposition. In 2001, however, such groups had been suppressed. Qaddafi thus no longer had to fear that acceding to U.S. demands would reveal his regime as weak and trigger a revolt. To the contrary, it only decreased the possibility the United States would impose regime change.

With regard to Qaddafi remaining in control of his own regime, the relevant question is how significant he anticipated his audience costs would be for backing down. As leader of a personalist regime, Qaddafi should have been less likely to be punished by the principals within his regime who might have been in position to remove him from power. Three reasons further mitigated Qaddafi's concern that his leadership might be challenged from within. First, his hold on power had improved substantially since the 1990s. Qaddafi had taken several measures to improve his personal safety. As a result, reports of coup attempts no longer surfaced. Second, the rise of militant Islamic organizations in the 1990s, some of which were backed by Al Qaeda, had caused Qaddafi to reverse his policies regarding terrorism. By 2002 Libya was publically cooperating with the United States over Al Qaeda, and what had been a root cause of conflict since the 1970s was no longer an issue. Qaddafi therefore did not have to concede this issue, which further reduced the audience costs he was likely to suffer.

Finally, Libya's WMD programs were now far less important to the regime. For three decades, Libya had been unsuccessful in its efforts to procure and produce nuclear weapons. Even A. Q. Khan's recent injections of nuclear enrichment and weapons technology had not improved the scientific and technical expertise required for Libya to develop a viable nuclear weapon. In addition, the high cost of maintaining a WMD program had been a continual source of tension within the regime since the mid-1970s. Given the high costs and lack of progress with its nuclear program, it is likely that those within Qaddafi's regime who viewed the nuclear program as an economic albatross might have embraced a change in policy. The combination of Qaddafi's firm control over his regime, his policy change over terrorism, and the expense and failure of Libya's nuclear weapons program reduced the audience costs Qaddafi was likely to suffer for conceding to U.S. demands.

In conclusion, the survival hypothesis correctly expected coercion to succeed, as neither the Libyan state, nor Qaddafi, nor his regime was likely endangered by Libya abandoning a WMD program that consisted only of chemical weapons. Other conventional explanations for bargaining failure and war do not fare as well in explaining the success of the United States.

Conventional Explanations for Bargaining Failure and War

The conventional explanations for war introduced in Chapter 2, that is, misperception and miscalculation, uncertainty and private information, and commitment issues, all provide reasons for why coercion should fail and states resort to war to solve their conflicts. These explanations do not, however, provide explanations for why coercion should succeed other than tautological arguments that, in successful cases, misperception, uncertainty, or commitment problems must not have been significant. For this crisis, it is most useful to analyze one of these conventional explanations by assessing how the United States avoided commitment problems with Libya. This is particularly perplexing because the United States had just demonstrated in the winter of 2002–2003 its propensity to increase its demands in reaction to Iraq's concessions, just as the logic of commitment problems would expect of a powerful challenger.

Given that the White House had ratcheted up its objectives to that of regime change even after Iraq disclosed that it had no WMD, how then was the administration able to avoid commitment problems with Libya? Four reasons help explain this puzzle. First, the United States had publicly acknowledged that Libyan regime change was not a policy objective. In the 1990s, while Iraqi regime change had become a U.S. objective, the Clinton administration made it clear that this was not the case for Libya. Further, after the September 11 attacks, Bush continued this policy excluding Libya from his "Axis of Evil."

Second, the order in which the White House sequenced negotiations incrementally built trust that the United States would make good on its promises. In August 2003, Libya agreed to settle the issue of the Pan Am bombing by acknowledging responsibility and paying compensation. In return, the United States, as promised, allowed the permanent lifting of suspended UN sanctions. This established a pattern of tit-for-tat concessions that ran counter to the logic of commitment problems. Qaddafi came to believe that, if he abandoned WMD, he could expect the United States to normalize relations as it had agreed to, rather than increase its demands, a commitment the Bush administration, in fact, fulfilled in June 2004.

Third, the fear that the United States might expand its objectives to include regime change in Libya, as it had in Iraq, had eased as the insurgency flared up in Iraq. By December 2003, when Qaddafi agreed to abandon Libya's WMD, the average number of daily attacks against coalition forces in Iraq had risen nearly tenfold since Bush's declaration of the end of major combat operations in May.[116] As the U.S. military was becoming increasingly enmeshed in its counterinsurgency operations in Iraq, it appeared less and less likely that the United States would be willing to open up a third front in Libya. The likelihood that the United States could credibly mount military threats to back up additional demands, as anticipated by the commitment problem explanation for war, waned with time.

And finally, the negotiations included Britain. This third-party involvement by a great power, even though an ally of the United States, increased the diplomatic costs had Bush later reneged on his promises.

CONCLUSION

This chapter assessed interstate asymmetric conflict between the United States and Libya from the time Ronald Reagan entered the White House in January 1981 until Muammar Qaddafi abandoned Libya's WMD in December 2003.

Three cases were considered during this time period: Libyan terrorism from 1981 to 1988, which resulted in the El Dorado Canyon air raid on April 15, 1986, but ended in coercion failure with the bombing of Pan Am Flight 103 over Lockerbie, Scotland; the extradition of two Libyan officials to stand trial for the Pan Am bombing, which ended as an economic coercion success when the United States agreed for a trial at The Hague; and, following September 11, 2001, Libyan WMD, a crisis that ended as a coercion success when Qaddafi conceded in December 2003.

The asymmetric coercion model provided a useful framework for analyzing all of these cases. In all three, the United States, as the great power, chose coercion over alternative strategies. In the Pan Am Flight 103 bombing case, the United States did not threaten force but did threaten and impose economic sanctions. This case expanded the scope conditions for the asymmetric coercion model to include cases for which economic sanctions replaced limited military force. Even without the threat of violence, the model did well in explaining the decisions made by the United States and Libya, as well as the final outcome of the case. In regard to the equilibrium conditions for the asymmetric coercion model, only in the El Dorado Canyon case did coercion fail and a coercion range not materialize. With the Pan Am bombing and Libyan WMD, however, a coercion range did exist, wherein the United States as the powerful challenger preferred coercion to brute force or accommodation and Libya as the weak target state preferred concession over resistance. Finally, the survival hypothesis expected all three conflicts to result in coercion success as the survival of the Libyan state, Qaddafi, and his regime were not threatened by making concessions. Only in the case of El Dorado Canyon was the survival hypothesis inaccurate. The stalemate was likely the result of Reagan miscalculating and not being able to maintain a credible threat of force, whereas Qaddafi proved reluctant to abandon his terrorism policies and reveal himself as being weak to those in his regime who might have plotted his overthrow.

The survival hypothesis did not fully explain the outcome of the El Dorado Canyon crisis, but it performed better than the conventional explanations for coercion failure and war, that is, misperception and miscalculation, uncertainty and private information, and commitment problems.

7 CONCLUSION

THE UNITED STATES HAS TIME and again threatened to use military force against much weaker states. Since World War II, the United States escalated asymmetric interstate conflicts into crises on thirty occasions; in two-thirds of these cases, it achieved its core foreign policy goals. U.S. leaders chose coercion over the alternative of brute force war in twenty-three of the cases (Table 1-1). Employing threats or using restricted force to convince a target to do its will, coercion has proven a popular military strategy. Yet U.S. efforts at coercion have succeeded only half the time (eleven of the twenty-three cases), far less than the five out of seven times the United States successfully seized its objectives by force.

This book has taken on the task of explaining not only why the United States so often chooses coercion but also why its efforts so often fail. Chapter 2 developed a theory of coercion specific to asymmetric interstate conflict and calculated equilibrium conditions and a coercion range whereby the United States, as the powerful challenger, prefers a negotiated settlement to war and the weak target state prefers concession to resistance. Chapter 3 provided an explanation for coercion failure based on the survival concerns of the weak state. States resist so long as they have the means to do so when concession threatens the survival of the state, the regime, or its leadership. This chapter also supplied an explanation for the follow-on question of why U.S. leaders would choose coercive strategies likely to fail. The United States is expected to conduct negotiations and/or obtain Security Council resolutions before it resorts to war. Indeed, such diplomatic efforts often sway allies and isolate opponents, thereby increasing the probability of victory and lessening the costs of war. Chapters 4 through 6 examined crises between the United States and Iraq, Serbia, and Libya through the lens of the asymmetric coercion model and evaluated the survival hypothesis against the conventional explanations for war.

The asymmetric coercion model assumes that, given the vast disparity in power, the United States has the latitude to decide whether to accommodate a weak target state or to escalate the conflict into a crisis by adopting a coercive or brute force strategy. Although the powerful state gets to decide whether to coerce, the weak state has the final say as to whether coercion will succeed or fail. Rationally, U.S. leaders should coerce only when they expect the targeted state will concede. This requires tailoring threats to demands and then credibly communicating intentions to the weak state. For example, after consulting President George W. Bush in September 2002, British Prime Minister Tony Blair demanded of Colonel Muammar Qaddafi that he abandon Libya's WMD. The United States and Britain then communicated the credibility of military action against Libya by invading Iraq and deposing Saddam Hussein. In response, Qaddafi agreed to renew negotiations, and, by December 2003, he had forsaken all WMD.

Although U.S. attempts at asymmetric coercion worked half the time, coercive diplomacy succeeded in only a third of the cases. Coercive diplomacy is a more restricted form of coercion for which threats are issued but force is not employed. The diminished effectiveness of coercive diplomacy is understandable for three reasons. First, a weak state often resists until the powerful challenger demonstrates its resolve by employing force. Second, great power leaders often learn from a weak state's recalcitrance that the initial coercive offer requires adjustment, with either a decrease in demands or an increase in threats. Finally, the leaders of weak regimes are concerned about the humiliation of concession. Though they may be willing to concede, they may prefer first to stand firm and fight, to receive a black eye before giving up in order to maintain a reputation for toughness. Examples of all three of these reasons were at play in U.S. coercive diplomatic efforts at Rambouillet, France, in February 1999, efforts that failed to convince Serbia's President Slobodan Milosevic to concede Kosovo. NATO's subsequent air campaign was initially ineffective. Over several weeks, however, the air strikes placed increasing pressure on Serbia. In May, the Clinton administration also began reducing its demands on Serbia. Russian troops would be allowed into Kosovo as peacekeepers, and all references to Kosovar independence were removed. Though coercive diplomacy failed, coercion succeeded in bringing Milosevic to concede in June 1999 after seventy-eight days of NATO air strikes. When addressing the Serbian people, he attempted to save face by arguing that they had stood firm in the face of NATO air power and had received a better peace agreement as a result.

As with the Kosovo crisis, the United States should theoretically continue with a coercive strategy only when it expects coercion to ultimately succeed. In reality, however, coercion often falls short, as when President Reagan could not convince Qaddafi to abandon his terrorism policies, which ultimately led to the bombing of Pan Am Flight 103. Why states cannot reach peaceful resolution to their conflicts remains a central question for international relations scholars. Unfortunately, the conventional nonrational and rational explanations for war, examined in Chapter 2, proved inadequate to explain the coercive outcomes for asymmetric interstate crises. Individuals, groups, and organizations do not always behave rationally in governing their states. Misperception of the capabilities and intentions of opponents can, in turn, result in miscalculation. In the crises detailed in this book, nonrational behavior is fundamental to understanding the onset of crises, the failure of coercive diplomacy, and the protraction of several conflicts. For example, Saddam Hussein misperceived the extreme imbalance of military power in the Persian Gulf in 1990–1991, which led to his army's decisive defeat. Likewise, in 1999 the Clinton administration miscalculated the resolve of Milosevic, expecting him to quickly cave in to demands over Kosovo as he had over Bosnia. U.S. leaders and their counterparts often make mistakes that result in violence and coercive diplomacy failure.

Nonrational explanations for war do not, however, explain the final outcomes of crises. Military force "provide[s] the stinging ice of reality," converging the perceptions of both the challenger and the target.[1] Though biases and misperceptions may cause a crisis and lead to the outbreak of fighting, the employment of violence yields information required for both sides to seek resolution. In addition, nonrational explanations for war have limited utility in anticipating the outcome of a new or ongoing crisis. States and their leaders exhibit nonrational behavior in all conflicts, yet only some crises result in coercion failure. Nonrational behavior does not provide ex ante indicators as to when coercion is likely to succeed or fail.

Closely related to nonrational causes of war is a rational explanation based on uncertainty and asymmetric information. This line of reasoning argues that uncertainty is prevalent in all conflicts. As a result, coercion fails when powerful challengers and weak targets arrive at different conclusions as to expected outcomes. If the two opponents would but share information, war could be avoided. Unfortunately, there exists an incentive to bluff, that is, to keep information private to improve one's negotiating position despite the risk of war this tactic entails.

Uncertainty clearly plays a major role in the crisis decision making of the United States and weak states alike. For example, ambiguity over NATO's resolve led Serbia to resist demands to concede Kosovo and to endure seventy-eight days of bombing. Over time, however, doubt diminished as NATO demonstrated its resolve to continue its air campaign. As with nonrational explanations for coercion failure, uncertainty and private information contribute to our understanding of the onset and duration of crises, but they do far less to help explain or anticipate their eventual outcomes.

A final structural explanation for war focuses on commitment problems resulting from the anarchic nature of the international system. Without a hegemonic power to enforce its agreements, the United States can renege on its promises. When a weak state then makes concessions, this can reveal the weak state as unresolved and further shift the relative balance of power. Although it is advantageous for the United States to reach an agreement, ex ante, once concessions reveal the target as weaker or less resolved than previously thought, the United States may be tempted to demand even more. Aware of this incentive, the weak state discounts the credibility of U.S. promises of future restraint.

Although the logic of commitment problems is engaging, two significant limitations emerge as to its utility in explaining crisis outcomes. First, as with the other conventional explanations for war, the commitment logic should be anticipated in all asymmetric crises, rendering the explanation unhelpful in predicting whether coercion is likely to succeed or fail when a new crisis arises. Second, in the cases examined in this book, there is scarce evidence of a weak power resisting because of commitment concerns. This was particularly striking in 1991, when Saddam conceded to the Soviet peace proposal only to have Bush issue a forty-eight-hour ultimatum, increasing U.S. demands for Iraq to effectively abandon its heavy weapons in Kuwait. At a point when commitment issues should have been paramount, Saddam voiced no such concern.

SURVIVAL: EXPLAINING WHY AND WHEN WEAK STATES
AND REGIMES RESIST

Opposed to the conventional reasoning for war, this book promotes an alternative explanation for coercion failure based on the survival concerns of weak states. In asymmetric conflicts the United States, as a great power, often has the capacity to threaten the survival of weak states. The more lopsided the balance of power, the more likely U.S. leaders will set demands that, if met, threaten the survival of the target state or its regime. Anticipating that concessions will only lead to this demise, leaders resist.

Chapter 3 outlined the conditions for state survival, that is, that states maintain domestic sovereignty over homeland territory and international sovereignty over foreign policy decision making. The advantage of the survival explanation as compared to the competing conventional explanations for war lies in the fact that coercive demands are typically known at the outset of a crisis. As such, the success or failure of a coercive strategy can be anticipated by the survival hypothesis, which expects a weak state to accede to demands unless concession threatens the survival of the state, the regime, or its leaders.

Survival concerns explain why the Bosnian Serbs would not concede territory until the Croat Muslim offensive placed them on the defensive in the summer of 1995. It also explains why Iraq and Libya both abandoned their WMD programs. Though the survival hypothesis correctly explained five of the seven coercive outcomes in this book, it incorrectly expected that Serbia would resist NATO over the issue of Kosovo. As it turned out, Serbia's concerns over its war-torn economy trumped its desires to retain its historic birthplace. The survival hypothesis also failed to anticipate the failure of the Reagan administration to maintain coercive pressure on Qaddafi over Libya's support of international terrorism during the 1980s. The survival hypothesis expects the powerful challenger to successfully implement its coercive strategies and therefore did not anticipate Reagan's inability to make credible threats after the El Dorado Canyon air raid. Though it does not explain the outcome for all crises, overall the survival hypothesis provides a better explanation for coercion failure than the conventional rational and nonrational explanations for war.

In analyzing cases of asymmetric coercion, it becomes apparent that the leaders of weak states are concerned not only with the survival of the state but also with their regime remaining in power. To incorporate this insight, the coercion model relaxed the unitary actor assumption to also consider the survival of the ruling regime and its leadership. Still, one question required additional consideration. Although survival concerns explain why weak states resist great powers, they do not explain why the United States would then intentionally adopt a coercive strategy that threatens a target's survival.

WHY THE UNITED STATES THREATENS THE SURVIVAL OF WEAK STATES

Rationally, the United States should understand that coercion is likely to fail when demands threaten the survival of a weak state, its regime, or its leaders. Only in cases in which weak states do not have the military capability to resist have U.S. leaders made such heavy demands and met with success. Helpless in the face of credible U.S. threats, a military junta conceded to regime change in 1994, as U.S. Air Force transport planes loaded with airborne troops approached Haitian air space.

Misperception, miscalculation, and uncertainty explain why the White House at times mistakenly makes demands that threaten a target state's survival. These explanations do not, however, explain why the United States would knowingly choose a coercive strategy to obtain high-level objectives of homeland territory or regime change. This book provides two related explanations for why a great power has an incentive to intentionally choose a coercive strategy likely to fail. First, it might calculate that it can achieve its foreign policy objectives only by brute force. Preparing for large-scale military operations takes time, however, and in the interim it may make sense to adopt a coercive strategy. For this to be a reasonable course of action requires that there be some uncertainty as to whether the weak state might in fact be coerced. Absent this uncertainty, the United States should not give up the element of surprise or divert its resources from its preparations for war. It is more likely that such an interim coercive strategy will be adopted if its implementation costs are low. This is particularly likely in denial strategies for which strikes against a target state's military may not only prepare the battlefield but also demonstrate the futility of enemy resistance. In 1990, the U.S.-led coalition imposed five months of economic sanctions followed by five weeks of air strikes in a combined effort to coerce and weaken the Iraqi Army in anticipation of the invasion of Kuwait.

A second and more pronounced incentive for the United States to choose coercion first is to reduce the costs of going to war and increase the probability of victory. In the modern world, international norms of behavior require states to first attempt to settle a dispute through negotiation or mediation before re-sorting to violence. Fighting without first making a civilized effort at peaceful resolution produces negative externalities. Such aggressive action can prompt a third party to intervene on behalf of the targeted state. War without negotia-tion might also damage relations with other states and threaten issues of even greater importance to the United States. In addition, U.S. leadership may be weakened politically for using brute force without first obtaining authoriza-tion from the Security Council. Going to the United Nations may also have the advantage of garnering allies and/or isolating the target state. For instance, to have a willing ally in Great Britain the George W. Bush administration made a final effort at obtaining a Security Council resolution authorizing force prior to its invasion of Iraq in 2003.

CRISIS SUMMARIES AND LESSONS TO BE LEARNED

U.S. leaders do on occasion intentionally choose coercive strategies likely to fail when doing so serves their broader purposes. The failure to coerce may be the outcome the United States seeks to justify the war it wants. This observa-tion, along with several other key insights into asymmetric coercion, is best summarized by briefly reexamining the findings from Chapters 4 through 6. Each crisis highlights lessons that, if properly digested by policy makers, could result in fewer crises and better decision making in the crises that do occur.

The Gulf War

The Gulf War is an excellent case for assessing three attributes of asymmetric coercion. First, the case tests the upper limits for the level of demands a pow-erful challenger can successfully pursue through coercion. When the U.S.-led coalition commenced air strikes on January 17, 1991, coercive diplomacy had failed to convince Saddam to concede to the demands set out in UN Security Council resolutions. Iraq's position did soften, however, when, after five weeks of air strikes and on the eve of the ground invasion, Saddam agreed to the Soviet-brokered peace proposal. Preempting this concession, President Bush issued a forty-eight-hour ultimatum for the complete withdrawal of Iraqi troops from Kuwait. Though Saddam could be coerced into giving up Kuwait, he would not be humiliated by ordering his army to abandon its weapons in a hasty retreat. The Iraqi Army was the source of his power and the only ob-stacle standing between U.S. forces and Baghdad. The fact that Saddam could not be coerced into making concessions that threatened his regime and the

Iraqi homeland supports the primary finding of this book: The leaders of weak states will resist when their survival is on the line.

Second, the Gulf War demonstrates why the United States may choose coercion even when it believes such a strategy is not likely to work. In October 1990, doubts over the ability to coerce Saddam through sanctions alone led Bush to approve plans for an invasion of Kuwait. The United States left sanctions in place, however, and then sought approval from the UN Security Council to conduct combat operations. This diplomatic effort succeeded in garnering international and domestic support while further alienating Iraq. The casus belli provided by the Security Council on November 29, 1990, required the United States to abide by the coercive strategy contained within the resolution which set a January 15, 1991 deadline.[2] The diplomatic and domestic political costs to be paid for invading any earlier were sufficient to prompt Bush to accept the deadline, despite the fact that coercive diplomacy was not likely to succeed. Nor did he even want coercion to work at this point, as the coalition had already deployed their forces. Successful negotiation would have left hundreds of thousands of coalition troops in the Middle East for an undetermined period of time. The price Bush had to pay for justifying an invasion was the risk that Saddam might find the terms acceptable and thus deny the administration one of its primary objectives, that of destroying the Iraqi Army.

Third, this case provides a strong test for commitment problems as an explanation for war. The U.S. military had sufficient deployed forces capable of invading Iraq. When Saddam agreed to the Soviet peace proposal, he was well aware of the capability of the United States to threaten the Iraqi homeland and his regime. Later, he would even claim victory on the basis that the coalition had not deposed him.[3] When Saddam attempted to concede, Bush increased the demands, just as the logic of commitment problems would expect. Yet, despite these optimal conditions for a credible commitment problem to exist, Saddam still tried to concede to the Soviet peace proposal and never voiced concern that the United States would only increase its demands in response.

Two related factors help explain why commitment problems did not play a prominent role in Saddam's decision to accept the Soviet proposal. First, the involvement of the Soviet Union in negotiations lessened Saddam's concern that the United States would go on to expand its demands. Mikhail Gorbachev placed his reputation on the line by negotiating a peace deal. The White House would have suffered significant diplomatic costs from the deterioration in its relationship with the Kremlin had Bush first agreed to the Soviet proposal, only to then renege once the Iraqi Army had withdrawn from Kuwait. The United States had vital national security interests at stake with the Soviet Union and wished to maintain a cooperative relationship. Second, the United States had formed a coalition that included Arab states. This coalition would likely have fractured had the United States invaded Iraq. The lesson to be drawn is that the involvement of another great power and the formation of a coalition decrease the likelihood that commitment problems will impede the resolution of a crisis.

The finding that commitment problems did not affect Saddam's decision making does not imply that they do not matter at all but underscores the fact that survival is a more pressing issue. A weak state worried about its own survival is not likely to be as concerned about how a concession now is likely to affect a potential crisis later. When death is at stake, the shadow of the future is short. Indeed, measures taken by a powerful challenger to alleviate commitment problems increase the chance of coercion success and may be even more significant for cases in which the weak state's survival is not threatened.

THE IRAQ WAR

The crisis leading up to the U.S. invasion of Iraq in 2003 supports the lessons from the Gulf War, but it also provides an additional insight into asymmetric coercion. First, just as his father had done prior to the Gulf War, George W. Bush felt obliged to go to the United Nations before invading Iraq. Unlike the Gulf War, however, this time Saddam cooperated with the United Nations. He allowed in weapons inspectors, thereby depriving the younger Bush the justification for the war he sought. This unintended coercive diplomacy success increased subsequent U.S. diplomatic costs for the war. The fact that Bush ordered the invasion, regardless, demonstrates the limits to which international norms can constrain a determined great power.

BOSNIA

The Bosnian Civil War provides two additional insights worthy of note. First, this crisis demonstrates the limits to the credibility of the threat of force when U.S. vital national interests are not at stake. President Clinton would not risk deploying ground troops for combat operations, a refusal that restricted the coercive leverage the United States could muster. Not until Croat and Muslim ground forces, supported by NATO airpower, threatened to overrun western

Bosnia did the Bosnian Serbs concede negotiating power to Serbian President Slobodan Milosevic.

Second, this is a case for which sanctions alone succeeded in changing a target's behavior by convincing Milosevic to withdraw his support for the Bosnian Serbs. Milosevic calculated that the negative impact of Serbia's decimated economy outweighed the benefits he garnered from supporting the Bosnian Serbs. Here, sanctions effectively changed Serbia's policy toward Bosnia, a critical factor in bringing the Bosnian Civil War to an end.

KOSOVO

The Kosovo crisis provides two insights into asymmetric coercion. First, it is the sole case for which the survival hypothesis produced a false positive, expecting failure when the outcome proved a success. Because Serbia viewed Kosovo as part of its homeland, the survival hypothesis expected Serbia to resist U.S. demands. Though Milosevic would not give up Kosovo without a fight, once his strategy to fragment the NATO alliance failed, he was no longer willing to endure the economic costs to Serbia of an extended air campaign. With its relatively small Serbian population, limited economy and material resources, and inconsequential geostrategic position, Kosovo was never critical to the survival of the Serbian state. The lesson is that not all homeland territory is the same, and states will resist to varying degrees based on the value of the territory relative to other vital issues at stake.

Second, Kosovo is a case in which the United States did not behave as a unitary actor. Coercive diplomacy failed, and the conflict escalated into violence, in part because the Clinton administration did not coordinate its demands and threats. Only after the Račak massacre in mid-January 1999 did Secretary of State Madeleine Albright finally convince the other principals of the Clinton administration to go along with her diplomatic initiative at Rambouillet, France. She demanded Serbia concede control over Kosovo or face NATO air strikes. Defense Secretary William Cohen and Chairman of the Joint Chiefs of Staff General Hugh Shelton stood fast, however, refusing to consider the use of ground troops over nonvital U.S. security interests. Coercive diplomacy failed when, in late March 1999, NATO commenced air strikes on Serbia. The initial threat from limited air strikes was not commensurate with the demand for Serbian homeland territory until the Washington summit in late April, when NATO's newfound resolve led to an escalation in the severity and timeframe for air strikes against Serbia. The not-so-surprising lesson is that coercive diplomacy may fail when there is a mismatch in demands and threats due to bureaucratic politics and the varying interests of key players and organizations within government.

EL DORADO CANYON

The crisis in the 1980s between the United States and Libya over Qaddafi's terrorism policies provides two more insights into asymmetric conflict. First, it is a case for which the Reagan administration combined coercive and brute force strategies. The White House made coercive demands of Libya to reverse its terrorism policies. The United States also conducted covert operations in an effort to topple his regime. The fact that Reagan chose not to attempt to coerce regime change supports the asymmetric coercion model's equilibrium conditions that a powerful challenger should not choose coercion when a target state is likely to resist.

Second, this case demonstrates the limits to coercion when the United States faces constraints on the level of force it is willing to employ. Given the domestic and international pressure on Reagan to abstain from using force, he authorized strikes in retaliation for terrorist attacks only when presented with undisputable evidence of Libyan involvement. Following the El Dorado Canyon air raid, Qaddafi undercut the United States' justification for further military action by reducing Libya's direct involvement in terrorist attacks, by shifting to covert support of terrorist groups, and by suppressing his anti-Western rhetoric. As a result, the crisis stalemated, and, although terrorist attacks by Libya decreased over the next two years, the raid proved only a short-term solution to a longer-term problem, as evidenced by the bombing of Pan Am Flight 103 in December 1988.

PAN AM FLIGHT 103

The bombing of Pan Am Flight 103 created a second case between the United States and Libya and generates three insights into asymmetric conflict. First, rather than the threat of military force, this case involved only the use of sanctions. The United States employed economic sanctions as a punishment strategy aimed at convincing Qaddafi to extradite the two Libyan officials suspected of the bombing. Economic sanctions eventually did influence Qaddafi in his decision to extradite, a concession he made to support his broader objective of normalizing relations with the United States and Western Europe.

The lesson is that, under certain conditions, sanctions can generate sufficient coercive pressure to convince a weak state to change its policies.

Second, although sanctions eventually produced a positive outcome for the United States, their use also had the real potential for blowback. Sanctions, coupled with depressed global crude prices, stagnated Libya's oil-dependent economy. Radical Islamic militant groups took advantage of the dismal economic environment to recruit Libya's disillusioned and unemployed youth. In one respect, this benefited the United States, as the threat these groups posed to his regime caused Qaddafi to reverse his proterrorism policies. Had the militants succeeded in overthrowing Qaddafi, as they went on to do in 2011, they could have posed a greater long-term threat to U.S. security interests. The United States would have belatedly achieved Reagan's objective for Libyan regime change only to find itself dealing with a country ruled by a radicalized Islamic regime with Al Qaeda ties.

Third, a key step in resolving this crisis was the 1998 Clinton administration renouncement of Libyan regime change as a U.S. foreign policy objective. For nearly two decades, Qaddafi had refused to make any concessions to the United States, a stance that changed only after the lifting of the threat of regime change. Subsequent Libyan concessions support the survival hypothesis in that, once survival was no longer an issue, Qaddafi could be convinced to make policy changes, even when demands were backed by sanctions alone.

WEAPONS OF MASS DESTRUCTION

The crisis over Libya's WMD is useful for examining two final aspects of coercion. First, the credibility of any threat of force the United States could leverage against Libya was enhanced by the U.S. invasion of Iraq. Following September 11, 2001, Qaddafi cooperated with U.S. intelligence against Al Qaeda. He also agreed to negotiations to settle Pan Am Flight 103. It was not until March 2003, however, that he finally agreed to negotiate over WMD. U.S. military action in Iraq demonstrated to Qaddafi that his WMD program did not increase Libya's security. On the contrary, not only did WMD fail to deter the United States, but they actually provided Bush with a casus belli, thus increasing the likelihood of U.S. aggression.

Second, the United States overcame a potential commitment problem caused by a lack of trust between the two countries. An acceptable level of trust developed from the involvement of Britain and a series of orchestrated diplomatic actions. Libya agreed to first announce the final settlement of Pan Am Flight 103, in return for which the White House publicly supported the permanent lifting of UN sanctions. The byproduct of both parties fulfilling their promises over this relatively minor issue not only increased the credibility of both states but also instilled a belief that further negotiations could be mutually beneficial. This set the conditions for bringing this two-decade conflict to a conclusion, as seen when Qaddafi agreed to abandon Libya's WMD in exchange for a normalization of relations with the United States.

IMPLICATIONS FOR U.S. FOREIGN POLICY

Having looked at each crisis in turn, one can draw several general foreign policy implications for asymmetric coercion. By far the most important application lies in recognizing the limits to which U.S. foreign policy objectives can be achieved through coercion. A primary finding affirms that high-level demands for homeland territory and regime change are likely to fail, as a target will not make concessions that threaten its survival. In their decision calculus, policy makers must carefully consider how demands will be perceived by the target, that is, not only whether demands threaten the survival of the target state but also whether the act of conceding alone will generate sufficient audience costs to threaten the survival of a weak state's regime or its leader. Policy makers should thus consider how the timing and content of the signals being sent will have an impact on a target's audience costs. Although public ultimatums and criminal indictments may play well domestically and internationally, such measures can also decrease the probability of coercion success.

At the lower range of foreign policy objectives, coercion may also fail when nonvital security interests are at stake. As demonstrated in both Bosnia and Kosovo, the ability of the United States to credibly threaten military force can be significantly degraded by its unwillingness to risk troops over nonvital interests. A further danger of escalating a conflict over limited objectives into a crisis is that, once an ultimatum has been issued, the prestige of the United States hangs in the balance. Although the original issue may have been a minor one, maintaining the reputation of the United States can prove vital to its leaders.[4] This can precipitate the expenditure of blood and treasure just to avoid losing the crisis, even if the costs outweigh the benefits of such a hollow victory. It is important for decision makers contemplating threats of force to realize that threats too often lead to the actual use of force; once blood is spilled, the stakes are raised.

A second implication from the case studies is that the United States can take steps to ameliorate potential commitment problems. The lack of trust is a key issue in an anarchic international environment, particularly for weak states that rightly fear great powers. Fortunately, U.S. leaders can take measures throughout a conflict to lessen a weak state's concern that conceding to demands today will lead only to further demands tomorrow. The formation of broad international coalitions, such as the coalition of the Gulf War, as well as the involvement of other great powers, such as Russia on behalf of Iraq and Serbia, increases the costs the United States would suffer for abrogating agreements. The implementation of incremental tit-for-tat concessions and conscious steps to reduce the audience costs born by the weak state's leadership are further measures that can be employed to mitigate commitment concerns. Such face-saving measures are particularly important when concessions do not threaten state survival but do threaten regime or leadership survival. In such cases, commitment problems may well be the determining factor between coercion success and failure.

Third, policy analysts should be alert as to the true intentions of a U.S. leader in employing coercion. Overreaching demands may merely be an attempt to mask preparations for war, as was the case for the elder Bush administration in the Gulf War. Demands may also be an effort to create a casus belli, as attempted by the younger Bush prior to the invasion of Iraq. Coercion failure does not necessarily mean foreign policy failure, and analysts should neither discount the effectiveness of coercion nor misconstrue the strategic intentions of leaders because of what on the surface appears a failure.

Finally, coercion success does not imply that the underlying conflict issues have been resolved. Coercion success does not change a weak state's preferences and may simply postpone a crisis over more important issues such as homeland territory or regime change. Coercion success that leaves a weak state's leader in power can lead to frustration for the United States and the eventual escalation of foreign policy objectives to regime change. This occurred in the late 1990s when U.S. foreign policy concerning Iraq shifting from containment to regime change. Likewise, key actors within the Clinton administration by early 1999 viewed Milosevic as the source of problems in the Balkans. Even Libya's concession of its WMD programs in 2003 did not satisfy the United States as it sought regime change in 2011 to solve what it perceived to be the root cause of conflict with Libya, that is, Qaddafi remaining in power.

THE FUTURE OF COERCION

Crises pitting the United States against much weaker states have been a common occurrence and are likely to remain so. It is difficult to predict the future for asymmetric conflicts, particularly because it is impossible to foresee such monumental occasions as the collapse of the Soviet Union or the 9/11 attacks and how these events can alter the balance of power in the international system or adjust U.S. perception as to what constitutes a vital security interest. Still, it is useful to briefly consider a recent asymmetric crisis between Libya and the United States and NATO played out against the backdrop of the Arab Spring of 2011. As the latest encounter between the United States and Libya, this crisis underscores a key finding of this book, that is, that weak states are not likely to concede to demands for regime change or homeland territory, so long as they have the means to resist. The United States must resort to war to meet its objectives in such cases. President Barack Obama identified regime change as a core objective in his March 28, 2011, address to the nation stating that Qaddafi "needed to step down from power."[5] NATO neutralized Qaddafi's military advantage by gaining air superiority and then dispersing his conventional forces with air strikes while simultaneously and covertly arming and training Libyan rebels in a civil war that ousted Qaddafi from power and led to his death. Given the limits to coercion, brute force will continue to be the most viable strategy available to the United States to achieve its most ambitious aims.

Second, coercive strategies involving sanctions and the threat of restricted force will continue to be employed as an interim stage in the lead-up to war. The United States will continue to impose sanctions on weak states even in cases when sanctions are likely to fail. In the case of Libya, the Security Council adopted resolutions imposing minimal sanctions against Libya and Qaddafi's regime. The UN Security Council adopted an arms embargo on Libya, which UN member states immediately began to violate by arming rebels. It imposed a travel ban on only a handful of Libyan officials, and it froze the financial assets of just a few members of Qaddafi's regime and a couple of Libyan corporations.[6] These sanctions were miniscule compared to the broad economic sanctions imposed against Serbia and Iraq during the 1990s. Targeted sanctions proved relatively easy to obtain Security Council approval for and cheap for the United States to implement. As a result, targeted sanctions are likely to continue to be used often. By the same token, because they place so little coercive pressure on weak states, they are not likely to succeed.

In addition to targeted sanctions, the United States will continue to coerce by using its asymmetric advantage in air power. For Libya, NATO leaders pushed through the Security Council a second resolution that Qaddafi end attacks against civilians and backed its demands by authorizing a no-fly zone over Libya. The actual implementation of the no-fly zone proved to be the first step in a brute force strategy. The no-fly zone quickly morphed into a no-drive zone while NATO concurrently armed and trained Libyan rebels. Though the conflict would last another five and a half months, NATO's decision to become actively involved in the fighting with only a limited authorization to use force still proved to be decisive to the outcome of the civil war.

Third, given the unprecedented power advantage the United States enjoys, asymmetric crises are likely to be recurrent phenomena. As the asymmetric coercion model demonstrates, the frequency with which the United States chooses coercive strategies and makes demands that threaten state and regime survival depends on the expected cost of fighting a brute force war. The Afghanistan and Iraq wars have informed American leaders of the true costs of such wars, and this has, at least for the time being, dampened U.S. enthusiasm for military intervention. Ceteris paribus, a rise in the expected cost for brute force war increases the likelihood the United States will instead choose a coercive strategy. Likewise, the intercession of great powers, whether China or Russia, or the involvement of regional powers would likewise increase the cost of U.S. intervention. Such powerful third parties could constrain the United States from using coercion or at least reduce the demands and threats made in those cases for which it does choose to coerce. Together, an increase in the expected cost of war and the intervention of great powers create incentives for the United States to accommodate a weak state and avoid altogether the threat of force.

· · ·

To conclude, this book has examined why the United States so often chooses to coerce weak states, why U.S. efforts at coercion so often fail, and why U.S. leaders on occasion choose to coerce even though they know coercion will fail. The hope is that, although we do not know where the next crisis will arise, whether in Syria, Iraq, Iran, or some other hot spot, a better understanding of the nature of asymmetric conflict and the requirements and limitations for successful coercive strategies will decrease the occurrence of avoidable wars and reduce the duration, pain, and suffering for those wars that cannot be avoided.

APPENDIXES

APPENDIX A

CODING U.S. CASES OF ASYMMETRIC COERCION

FOLLOWING ARE SUMMARIES OF SEVENTEEN CASES of U.S. asymmetric coercion listed in Table 1-1, U.S. Asymmetric Interstate Crises: 1950–2011, in Chapter 1. Not included are the six crises analyzed in detail in Chapters 4 through 6, between the United States and Iraq, Serbia, and Libya, respectively. The seven cases in which the United States chose a brute force strategy are also not considered here.[1]

1964: NORTH VIETNAM–GULF OF TONKIN[2]

The United States attempted to coerce North Vietnam into discontinuing its support of the Viet Cong in South Vietnam. After the dramatic defeat of France at Dienbienphu in 1954, Vietnam was divided into the communist North, led by Ho Chi Minh, and a U.S.-supported South led by Ngo Dinh Diem. The overall U.S. objective in Southeast Asia was to halt the spread of communism with a stable, noncommunist, and independent South Vietnam.[3] By 1963, the Kennedy administration, having lost faith in Diem, supported a military coup that destabilized an already weak South Vietnamese government. The North Vietnamese took this opportunity to expand the insurgency. By the spring of 1964, with the South Vietnamese government in disarray, the new Johnson administration shifted its focus toward the North with covert operations aimed at convincing Hanoi to stop providing troops, supplies, and guidance to the Viet Cong.

On August 2 and 4, 1964, the U.S. Navy destroyer *Maddox*, operating in the Gulf of Tonkin, reported being attacked by North Vietnamese torpedo boats. The United States retaliated with air strikes on a North Vietnamese naval base and nearby oil storage facilities. On August 7, the United States escalated the conflict into a crisis when Congress passed the Gulf of Tonkin Resolution authorizing conventional military force against North Vietnam.

Two factors, however, prevented the United States from maintaining a credible threat of military force. First, President Johnson had gained popular support for his reaction to the Gulf of Tonkin incident, and he chose not to risk further military action until after the November presidential elections. Second, General Nguyen Khan, the new leader of South Vietnam, exploited the Gulf of Tonkin incident by attempting to assume near-dictatorial powers. This led to widespread protests against him in Saigon, which resulted in Khan's resignation. With South Vietnam leaderless, the United States opted not to take further military action until Saigon's government had stabilized.[4]

The Gulf of Tonkin crisis ended as a foreign policy, coercion, and coercive diplomacy failure for the United States. The North Vietnamese continued to support the Viet Cong who, in turn, stepped up its attacks, including the November 1, 1964, attack on U.S. forces at the Bien Hoa Air Base.[5]

1965: NORTH VIETNAM—PLEIKU/ROLLING THUNDER[6]

The United States continued its effort to compel North Vietnam to discontinue its support of the Viet Cong in South Vietnam. The political situation within South Vietnam deteriorated rapidly over the winter of 1964–1965, with a third military coup taking place in Saigon in late January. On February 6, a Viet Cong raid against U.S. Army barracks at Pleiku killed nine soldiers; in response, the United States escalated the conflict into a second crisis with North Vietnam.

For a year, the Johnson administration had been considering air strikes against North Vietnam to provide breathing room for South Vietnam to sort out its political situation. The White House used the Pleiku incident as a trigger to initiate Rolling Thunder, a graduated air campaign designed to compel the North Vietnamese to withdraw its support of the South Vietnamese insurgency. Johnson's advisors, however, split over assessments as to whether air strikes alone could convince Hanoi to abandon its nationalist goal of a reunified Vietnam. Even supporters of Rolling Thunder expected it to take months before sufficient pressure would compel Hanoi.[7] To stabilize the situation in South Vietnam, the United States simultaneously commenced the deployment of combat ground forces to South Vietnam. Troops levels would grow to nearly half a million by January 1968.

Rolling Thunder commenced in earnest in March 1965 with Johnson approving all targets, beginning with military targets in the North and the interdiction of roads leading to the South. While the U.S. Air Force and Navy gradually escalated air strikes, the North Vietnamese ignored American demands by continuing to send troops and material southward. In July, Johnson sent Secretary of Defense Robert McNamara to Saigon; he returned with a gloomy assessment. More emphasis was placed on supporting ground operations in South Vietnam, though air strikes on North Vietnam continued. Johnson approved more targets, and, in the spring of 1966, the air campaign shifted to attacking North Vietnamese POL (petroleum oil and lubricant) sites. Throughout the fall of 1966 and into the spring of 1967, air strikes targeted North Vietnamese industry and electric power plants. The Rolling Thunder target list expanded further in October 1967 to include previously restricted targets in and near Hanoi.[8]

By 1968, three years of U.S. air strikes had failed to compel North Vietnam from supporting the Viet Cong insurgency. American foreign policy and coercion failed in this crisis, which concluded with the commencement of a third crisis after the North Vietnamese launched the Tet Offensive on January 30, 1968. Coercive diplomacy is nonapplicable for this case because the crisis occurred during an ongoing war with military force already being employed when it commenced.

1968: NORTH VIETNAM—TET OFFENSIVE[9]

As the Air Force and Navy commenced the Rolling Thunder air campaign in 1965, the commander of U.S. forces in South Vietnam, Army General William Westmoreland, formulated a separate coercive attrition strategy based on search-and-destroy missions against Viet Cong irregular units along with North Vietnamese conventional forces operating in South Vietnam. The objective was to compel North Vietnam to withdraw its support for the Viet Cong insurgency. The United States incrementally deployed combat forces, who numbered half a million in South Vietnam by January 1968.

Until 1967 North Vietnamese forces refused large-scale conventional engagements against the U.S. Army. Over the winter of 1967–1968, however, the North Vietnamese attacked small towns and American outposts throughout South Vietnam to draw ground forces into battle. In mid-January U.S. Marines engaged North Vietnamese forces in a large conventional battle at Khe Sanh. With U.S. forces drawn out of the cities, the North Vietnamese launched the Tet Offensive, simultaneously attacking major cities and provincial capitals throughout South Vietnam.

Though surprised, U.S. and South Vietnamese forces recovered and organized a strong defense. The Viet Cong and North Vietnamese units suffered heavy losses as Westmoreland finally got the attritional battles and operational victory he had been seeking. It was the United States, however, and not the North Vietnamese, that was coerced in this case. After Tet, the American public lost faith in Johnson's handling of the Vietnam War. Johnson rejected the deployment of more ground forces and, on March 31, 1968, announced a bombing halt over most of North Vietnam. He also announced that he would not be seeking reelection. The Tet offensive crisis ended as a foreign policy and coercion failure for the United States. Coercive diplomacy is nonapplicable for this case because it occurred during an ongoing war with military force already being employed when the crisis commenced.

1970: SYRIA—BLACK SEPTEMBER[10]

Following months of unrest in 1970, on September 1, Palestinian guerillas attempted to assassinate Jordan's King Hussein and on September 6 commandeered three American, Swiss, and British airliners to a remote airfield near Amman. The hijackers demanded the release of over a thousand Palestinians held by the British, Germans, Swiss, and Israelis.[11] The United States pressed Britain, Germany, and Switzerland to refuse the demands of the hijackers.[12] Within the week, the hijackers released most of the passengers but dramatically blew up the empty airliners. The United States then deployed a squadron of F-4s and C-130s to Turkey, repositioned a carrier battle group and troop carriers off the coast of Jordan, and ordered the mobilization of the 82nd Airborne Division at Fort Bragg.

King Hussein launched a major counteroffensive against the Palestinians residing in Jordan. He formed a military government, declared martial law, and deployed security forces into the Palestinian-controlled regions. In response, on September 19 the Syrians, who supported the Palestinians, sent armored units across the border into Jordan.[13] At this point the United States escalated the conflict into a crisis by demanding that Syria withdraw its forces. President Richard Nixon encouraged Israel to deploy fighter squadrons to the north to threaten Syrian forces. Nixon also assured Israel that American forces in the Mediterranean would deter Egypt from attacking Israel's now exposed rear.[14] On the diplomatic front, National Security Advisor Henry Kissinger began pressuring the Soviets to convince the Syrians to back down.

Meanwhile, Syria's lead armor units became bogged down under fire from the small Jordanian Army and Air Force. The ground forces required air support, but Syria decided not to commit its air force out of concern that Israeli fighters would attack. The crisis abated on September 22, when Syria withdrew its armor from Jordan. This crisis proved a foreign policy, coercion, and coercive diplomacy success for the United States, which, with the assistance of Israel and the Soviet Union, compelled the Syrians to withdraw from Jordan.

1972: NORTH VIETNAM—VIETNAM PORTS
MINING/LINEBACKER I[15]

By May 1971, the U.S. foreign policy objectives for Vietnam had been reduced from an effort to maintain a secure and independent, noncommunist South Vietnam to a peace agreement that would allow the United States to withdraw from South Vietnam. National Security Advisor Henry Kissinger secretly offered the North Vietnamese a withdrawal of U.S. troops in exchange for the release of American POWs and a promise that Hanoi would discontinue its infiltration of the South. These private negotiations finally broke down, however, over North Vietnam's insistence that South Vietnam's President Nguyen Van Thieu first be removed from office before Hanoi would sign a peace deal. Although Kissinger failed to reach an agreement with the North, by the end of 1971 he had succeeded in significantly improving relations with the Soviet Union and China.

In March 1972, the North Vietnamese launched the Easter Offensive deploying fourteen divisions (120,000 troops) in a massive conventional invasion of the South. By then, as a result of Vietnamization (handing over combat responsibilities to South Vietnamese forces), the U.S. military had fewer than 70,000 ground forces widely dispersed throughout South Vietnam. Nixon refused to deploy additional troops but did commit U.S. air power to support the South Vietnamese Army. In May he initiated the Linebacker I air campaign, aimed against North Vietnamese military logistics and transportation targets. The escalation in operations included tactical aircraft with laser-guided bombs and the mining of Haiphong harbor. Meanwhile, U.S. leaders pressured the Soviets and Chinese to convince the North Vietnamese to seek a diplomatic solution.

U.S. air power played a pivotal role in frustrating the North's designs in South Vietnam. North Vietnam's decision to employ heavy conventional forces left them vulnerable to air attack. By September the defeat of their Army divisions by the South Vietnamese army, supported by U.S. air power, convinced Hanoi to negotiate in earnest. In mid-October, the Americans and North Vietnamese reached terms, though a final peace agreement would be delayed by a reluctant South Vietnamese President Thieu until January 1973.

Overall, the United States achieved its foreign policy objectives by halting the Easter Offensive and compelling the North Vietnamese to settle on U.S. terms for a face-saving withdrawal of U.S. forces. Though a coercion success, coercive diplomacy is nonapplicable for this case because it occurred during an ongoing war with military force already being employed when the crisis commenced.

1972 NORTH VIETNAM—CHRISTMAS BOMBING/LINEBACKER II[16]

In October 1972 South Vietnam's President Nguyen Van Thieu blocked the signing of a peace agreement negotiated between North Vietnam and the United States. To assuage Thieu's concerns, Nixon promised to provide Saigon with billions of dollars in military aid, along with assurances that the United States would punish North Vietnam for any infractions to a peace treaty.

After weeks of additional negotiations between Kissinger and North Vietnam's negotiator Le Doc Tho, a frustrated Nixon terminated the talks in mid-December and ordered a second round of bombings. Linebacker II, also dubbed the Christmas bombings, consisted of eleven nights of air strikes by B-52s on rail yards, power plants, and airfields in Hanoi and Haiphong Harbor. Nixon intended Linebacker II to compel the

North Vietnamese to sign a peace deal and to demonstrate to Thieu that the United States would remain committed to South Vietnam even after the withdrawal.

By the tenth night of bombing, the North Vietnamese had run out of surface-to-air missiles to defend their skies. On December 28, Hanoi sent a message to Washington calling for a resumption of talks. Nixon ordered a bombing halt the next night, and on January 27, 1973, the United States and North Vietnam signed a peace agreement. U.S. POWs were repatriated, and Thieu stayed in power. In return, the United States stopped its bombing and withdrew all of its military forces. In addition, deployed North Vietnamese troops remained in South Vietnam.

Nixon finally succeeded in achieving his foreign policy objective of withdrawing the United States from Vietnam. U.S. air power had compelled the North Vietnamese to return to Paris and sign a peace agreement they had previously agreed to in October. Nixon further compelled Thieu to sign on by threatening to withdraw all U.S. aid. Coercive diplomacy is nonapplicable for this case because it occurred during an ongoing war with military force already being employed when the crisis commenced.

1975: CAMBODIA—MAYAGUEZ[17]

In April 1975 U.S. personnel evacuated Phnom Penh, Cambodia, and Saigon as South Vietnam fell to the North Vietnamese. Two weeks later, on May 12, a Cambodian Khmer Rouge gunship seized the *Mayaguez*, a U.S.-flagged merchant ship with a crew of forty, as it steamed through the Gulf of Siam.[18] President Gerald Ford escalated the conflict to a crisis by labeling the act as piracy and threatening serious consequences should the ship not be immediately released.[19] The United States diverted two aircraft carrier battle groups to the South China Sea, and 1,100 Marines deployed from Okinawa to Thailand. On the morning of May 13, a U.S. Navy P-3 surveillance plane located the *Mayaguez* anchored near the island of Koh Tang, thirty miles off the Cambodian coast. The P-3 observed the crew being taken from the ship onto two small fishing boats, and, in an effort to prevent the Khmer Rouge from taking their hostages to the mainland, U.S. Navy and Air Force jets flew low-altitude passes and strafed several boats, sinking three. Unfortunately, the aircraft lost track of the fishing vessels with the hostages.

On the morning of May 15, the Khmer Rouge broadcast on Phnom Penh Radio that the U.S. ship had been ordered to depart Cambodian waters. The *Mayaguez* crew was then taken aboard two small boats back to their anchored ship. The announcement was made, however, as 130 Marines assaulted Koh Tang Island, where the United States had mistakenly assessed the hostages to have been taken. The assault force met resistance from a hundred armed Cambodian soldiers on the island. Of the eight U.S. helicopters involved in the assault, three were shot down and four were so damaged that they were forced to abort. Meanwhile, Marines aboard three helicopters success-fully retook the abandoned *Mayaguez*, and, two hours later, the Cambodians delivered the Mayaguez crew to a nearby U.S. destroyer. It would take another ten hours to extract the Marines from Koh Tang Island.[20] A total of forty-one American military members were killed, eighteen in the attacks on Koh Tang Island and another twenty-three U.S. airmen who had died the day before the assault when a USAF helicopter crashed enroute from Thailand.

Despite these high military casualties, this crisis is coded a foreign policy success because the U.S. succeeded in gaining the release of both crew and ship, thus avoid-ing an extended hostage crisis. Coercion succeeded as the Khmer Rouge released the crew under the impending threat of attack. Coercive diplomacy, however, failed as the United States employed military force in the Marine's assault on Koh Tang and in simultaneous air strikes on a Cambodian air and naval base.

1976: NORTH KOREA—POPLAR TREE[21]

On August 18, 1976, a fifteen-man team consisting of South Korean workers protected by a U.S.–ROK (Republic of Korea) security detail set out to trim back a poplar tree. The tree, situated at the eastern end of the "Bridge of No Return," obstructed the view of a UN Checkpoint in the Joint Security Area of the demilitarized zone at Panmun-jom, Korea. In response, thirty North Korean soldiers armed with clubs and axes sur-rounded the team and, in the ensuing scuffle, killed two U.S. Army officers assigned to the detail.[22]

The United States initiated a crisis as a result of this incident. By that afternoon, the Washington Special Action Group, chaired by Secretary of State Henry Kissinger, convened to review the situation. Kissinger recommended to President Ford that he raise the alert status for military forces in the region. He further ordered the deploy-ment of an F-4 and an F-111 fighter squadron as well as a U.S. Navy carrier battle group to Korea.

On August 19, U.S. and North Korean officials met at Panmunjom, where the U.S. delegation read a prepared statement accusing the North Koreans of deliberate murder. The North Koreans dismissed the accusation of a premeditated attack, claim-ing that the UN detail had initiated the confrontation. In public statements, the State Department demanded that North Korea accept responsibility for the killings, make assurances they would not happen again, and punish those responsible.[23]

On August 20 the United States and South Korea executed Operation Paul Bun-yan, simultaneously chopping down the poplar tree while conducting a massive show of force, including B-52s from Guam escorted along the DMZ by U.S. and ROK fight-ers. In addition, an airborne infantry unit rode in helicopters, joined by Cobra attack helicopters.

An hour after the termination of Paul Bunyan, the North Korean senior representative at Panmunjom delivered a message from North Korea's leader Kim Il-sung. This was the first personal message from him in the history of the armistice. In the note, Kim Il-sung did not accept responsibility for the killings, nor did he indicate those responsible would be punished. He did, however, indicate regret that the incident had taken place and urged no further provocations by either side. The crisis deescalated when the United States accepted Kim Il-sung's reply and, in subsequent meetings in Panmunjom from August 25 to September 6, the Americans, South Koreans, and North Koreans adopted a new agreement establishing a military demarcation line within the Joint Security Area.

This crisis was a foreign policy success for the United States. It stabilized tensions between America and North Korea and led to a new Joint Security Area agreement to reduce further provocation. U.S. allies approved of the actions, and the Soviets and Chinese showed restraint. Though Kim Il-sung did not meet all three State Department demands, the White House accepted the letter as an apology and, as such, a coercion and a coercive diplomacy success.

1981: LIBYA—GULF OF SIDRA[24]

In January 1981, President Ronald Reagan entered the White House promising to restore America's military power and prestige. Libya provided low-hanging fruit for a more confrontational U.S. foreign policy as an outspoken Muammar Qaddafi supported international terrorist groups. Libya also actively sought nuclear and chemical weapons programs.[25] Reagan demanded Libya change its policies, although privately the White House did not believe such changes were likely.[26]

The United States provoked a crisis by conducting naval exercises off the coast of Libya designed to elicit a response from Qaddafi by challenging Libya's territorial claims to the Gulf of Sidra.[27] A military confrontation ensued on August 19 when U.S. Navy F-14 fighters operating south of Qaddafi's "line of death" downed two Libyan jets after being firing on by the Libyan fighters.[28] Qaddafi threatened but did not continue to respond directly to U.S. military actions.

A month later, the Reagan administration leaked to the press that it had intelligence of Qaddafi threatening to assassinate Reagan in retaliation for the incident.[29] The U.S. recalled American citizens from Libya in December 1981 and placed a unilateral boycott on Libyan oil. Although the administration attempted to elicit international support for sanctions against Libya, the unwillingness of the White House to release evidence of the alleged assassination attempt undermined these efforts. As a result, when Reagan announced additional sanctions against Libya in February and again in March 1982, U.S. allies balked at going along with the policy.[30]

The crisis concluded as a foreign policy, coercion, and coercive diplomacy failure as Libya neither changed its policies with regard to international terrorism nor abandoned its WMD programs.

1988: NICARAGUA—CONTRAS[31]

In 1981, President Reagan denounced Nicaragua's leftist Sandinista government for supporting rebel forces in El Salvador. The United States initiated a crisis by adopting a coercive strategy aimed at Sandinista regime change. Later that year, the White House canceled a multibillion dollar economic aid package to Nicaragua and authorized the CIA to assist the Contra rebels in their effort to overthrow the Sandinistas.

In an effort to defeat the insurgency in Nicaragua the Sandinistas announced a partial amnesty for the Contras in 1983 and made plans for a Nicaraguan national election. The Reagan administration rejected these conciliatory gestures and continued its covert activities. In 1984, after discovering that the CIA had been ordered to mine Nicaraguan harbors, Congress undermined the White House by passing the Boland Amendment to ban further aid to the Contras. Undeterred, the Reagan administration responded with a clandestine operation run from within the staff of the National Security Council, whereby it illegally sold arms to Iran and funneled the revenues to the Contras.

In the 1984 national election the Sandinistas won a majority of the vote. Reagan, however, denounced the results and, in 1985, imposed a trade embargo on Nicaragua and blocked loans from the World Bank and International Monetary Fund. Ongoing negotiations between Nicaragua and the State Department collapsed. Relations between the two countries reached their nadir in June 1986 when Congress removed its moratorium on Contra aid. Meanwhile, the Nicaraguans attempted to overcome its economic problems and settle local disputes by negotiating the Esquipulas II (Aris) Treaty, ratified in August 1987, with Costa Rica, Guatemala, Honduras, and El Salvador. The treaty called for economic cooperation and a framework for conflict resolution.[32]

Even with the treaty, by the end of 1987 the economic pressure of U.S. sanctions, coupled with the costs of fighting the Contras, drove Nicaragua into an unsustainable level of government spending, followed by hyperinflation and a severe recession. By the beginning of 1988, the Contras seemed to be gaining momentum, both politically and militarily. Nicaragua's Sandinista President Daniel Ortega agreed in January to begin cease-fire talks with the Contras. To keep pressure on the Sandinistas, the Reagan administration announced in early February that it would seek a $270 million aid package for the Contras.[33] In the House of Representatives opponents narrowly defeated the bill, however, and by the end of February, U.S. aid to the Contras had dried up.

With the Sandinistas now gaining the upper hand, they launched an offensive on March 6, 1988, against Contra camps within Honduras. On March 15, the Honduran president appealed for help, and the United States responded by dispatching 3,200 troops to the region. On March 17, Honduran fighter jets, provided and supplied by the United States, bombed Nicaraguan ground troops, compelling them to withdraw.

At the end of March 1988, the crisis deescalated when the Sandinistas agreed to a sixty-day cease-fire agreement, later extended until October 1989. The Contra movement, now engulfed in bitter infighting, collapsed politically and militarily from the loss of American financial support. The threat of U.S. military force had an immediate impact in convincing the Sandinistas to withdraw from Honduras and to negotiate a cease-fire with the Contras. The Reagan administration, however, did not achieve its core foreign policy objective of Nicaraguan regime change as its efforts at coercion and coercive diplomacy failed.

1993: NORTH KOREA—NUCLEAR CRISIS[34]

In 1982, U.S. satellites photographed the construction of North Korea's Yongbyon nuclear facility. In 1985, the Soviet Union agreed to supply four light water reactors on the condition that North Korea sign the Non-Proliferation Treaty (NPT). By December 1988, Pyongyang realized that the Soviet reactors would not be forthcoming and therefore refused the International Atomic Energy Agency (IAEA) inspections required by the NPT. In 1991, however, North Korea pledged not to pursue nuclear weapons and to allow in the IAEA, provided that the United States remove all its nuclear weapons from the Korean peninsula and cancel the upcoming U.S.–Republic of Korea (ROK) Team Spirit military exercise scheduled for 1992.

Tensions with North Korea soon returned, however, with the ROK and U.S. announcement that they would hold the exercise in 1993, and the IAEA demanded inspections of a previously undisclosed storage site at Yongbyon. When Team Spirit commenced in 1993 as scheduled, North Korea raised its military alert status and, on March 12, announced its intention to withdraw from the NPT. The Clinton administration threatened sanctions, but, after extensive talks, North Korea agreed to suspend its decision to withdraw from the NPT and to ensure safeguards against weaponization by allowing continued inspections. U.S. and North Korean leaders then issued mutual assurances not to use force against each other and to maintain a nuclear-free Korean peninsula.[35]

In the fall of 1993, however, negotiations between North Korea and the IAEA collapsed. Over the winter, the Americans escalated the conflict into a crisis by again threatening sanctions; in January, news leaked of the U.S. plan to deploy Patriot antiballistic missiles to South Korea and to conduct Team Spirit in November of 1994. Relations with the United States and North Korea reached a low point in the spring of 1994 when North Korea suspended IAEA inspections in March and then announced in April that fuel rods would be unloaded from the Yongbyon reactor. Once processed, this fuel could provide enough plutonium to produce five plutonium warheads.

On June 2 the IAEA announced that the unloading of the fuel rods had destroyed evidence of how much plutonium had been extracted from Yongbyon's reactor core. In response, North Korea announced its intentions to withdraw from the IAEA. The crisis escalated further as North and South Korea increased their military alert postures, and the Clinton administration contemplated taking military action and evacuating U.S. citizens from Korea.

On June 15, former President Jimmy Carter arrived in Pyongyang for a meeting with Kim Il-sung. The next day Carter announced on CNN that Kim had agreed to allow the IAEA to resume inspections in exchange for new light-water reactors and the United States discontinuing efforts at the UN Security Council for a resolution to impose sanctions. On June 20, the White House followed up Carter's visit by extending the offer to resume talks to replace the North Korean reactors if the North froze its nuclear weapons program and complied with the NPT by allowing the resumption of inspections. The crisis began to wind down when North Korea agreed to this proposal two days later, although final negotiations would drag on into October 1994.

The United States achieved its limited foreign policy objective of freezing North Korea's nuclear weapons program at Yongbyon. Coercion and coercive diplomacy succeeded through a combination of threats of force and sanctions, along with the inducement of light water reactors and a presidential assurance that the United States would not use force against North Korea.

In a final note, though this case is coded as a coercion success, the United States did not compel North Korea to permanently abandon its nuclear aspirations. With tensions again rising between North and South, due in part to a delay by the United States in delivering on its promise of the light water reactors, evidence of a clandestine centrifuge program in North Korea to enrich uranium emerged in 1999.

1994: HAITI—MILITARY REGIME[36]

On December 16, 1990, Haiti elected Jean Bertrand Aristide, a Roman Catholic priest, as president in its first democratic election. On September 30, 1991, a military coup, led by General Raoul Cedras, overthrew Aristide. By July 1994 the junta's repression of the Haitian population had produced tens of thousands of refugees, setting out in makeshift boats in an attempt to reach the U.S. mainland.

The refugee situation triggered a crisis for the United States and prompted the Clinton administration to adopt a foreign policy objective of Haitian regime change. The Americans worked through the UN Security Council to pass a resolution on July 31, demanding the return of Aristide to power and authorizing the use of "all necessary means to facilitate the departure from Haiti of the military leadership."[37]

By mid-September 1994, the White House released its plans for 20,000 troops to occupy Haiti. On September 16, Clinton agreed for former President Jimmy Carter, former Chairman of the Joint Chiefs Colin Powell, and Senator Sam Nunn to go to Haiti and broker terms for how the junta would depart Haiti and U.S. forces would peacefully occupy Haiti. On September 18, with U.S. Air Force transport planes enroute carrying elements of the 82nd airborne, Cedras agreed to exile. The next day, U.S. soldiers arrived in Haiti, and Aristide returned to power on October 15, 1994. The United States resolved the crisis as a foreign policy, coercion, and coercive diplomacy success for achieving its core objective of regime change.

1994: IRAQ—TROOP DEPLOYMENT KUWAIT[38]

By 1993, Iraq had suffered through three years of economic sanctions imposed on it after its invasion of Kuwait. Inflation spiraled out of control, leaving most Iraqis dependent on meager government food rations. In November 1993, Saddam Hussein reversed his policies on weapons of mass destruction and began to cooperate with UNSCOM (UN Special Commission) inspectors.[39] By the following summer, UNSCOM had destroyed all known chemical and nuclear weapons and production equipment.[40] As such, Saddam had been counting on a positive report in September and a recommendation from UNSCOM's chief to have the sanctions lifted. At this point, Iraq had exhausted its foreign currency reserves purchasing imported foodstuffs and halved its already scant rations for its population.[41] The UNSCOM chief instead recommended an extension of sanctions for another six months to ensure continued Iraqi compliance.[42] In response, Saddam mobilized two Republican Guard armored divisions consisting of 14,000 troops and deployed them to the Kuwaiti border.[43] This action coincided with an address by the Iraqi deputy prime minister to the UN General Assembly on October 6 in which he demanded "a change of this unjust and illegitimate situation as soon as possible."[44]

That same day intelligence sources discovered the movement of Iraqi forces to the Kuwaiti border, prompting U.S. officials to escalate the conflict into a crisis. The Pentagon responded immediately, repositioning forces in the region and announcing the deployment of thousands of troops and hundreds of additional warplanes. The United States issued coercive demands that Iraq not invade Kuwait and that the Republican Guard units return to their barracks. The Americans backed these demands with the threat of air and cruise missile strikes against Baghdad.

On October 10, Iraq announced the repositioning of its troops with the claim that they had only been conducting exercises. By October 17, U.S. intelligence confirmed that the units had returned to their base. This crisis concluded as a foreign policy, coercion, and coercive diplomacy success for the United States. Not only did the United States achieve its core foreign policy objectives with only the threat of military force, but Iraq went on to make a further concession to UN resolutions. On

November 10, with the encouragement of the Russians, Saddam Hussein formally recognized "the sovereignty of the State of Kuwait, its territorial integrity and political independence."[45]

1997: IRAQ—UNSCOM/OPERATION DESERT FOX[46]

Following the 1994 Kuwaiti border crisis, Iraq resorted to cheat-and-retreat tactics whereby the Iraqi government would prevent UNSCOM inspectors from having access to suspected weapons sites only to back down when threatened by air strikes.[47] Following the receipt of a scathing report by the executive chairman of UNSCOM on Iraqi intransigence, the Security Council passed two resolutions in October and November 1997. These demanded Iraq fully cooperate with inspectors, giving "immediate, unconditional and unrestricted access" to the sites and imposed travel bans on Iraqi officials who impeded inspections.[48]

Iraq responded to the resolutions on November 13, 1997, by expelling U.S. members of UNSCOM. The White House escalated the conflict into a crisis when President Bill Clinton announced the deployment of military aircraft into the region, including an additional carrier battle group and B-1 bombers.[49] Russia's Foreign Minister Yevgeny Primakov quickly brokered a deal for Iraq to allow U.S. inspectors back into country as long as UNSCOM respected Iraqi sovereignty. Within weeks, however, Iraq had again placed restrictions on UNSCOM. Iraq expelled American UNSCOM team leader Scott Ritter in January 1998 for reporting that Iraq had tested chemical and biological weapons on prisoners. Tensions remained high until late February when UN Secretary-General Kofi Annan traveled to Iraq and negotiated an understanding for how inspections would proceed. The Security Council endorsed the agreement.[50]

In July, Iraq refused to provide a copy of a list of chemical weapons expended during the Iran–Iraq war, and it formally suspended cooperation with UNSCOM inspectors in August, though it continued to allow monitoring equipment to remain in place.[51] In September, the Security Council unanimously condemned Iraq's decision and suspended the periodic review on lifting economic sanctions until Iraq reversed its decision.[52]

A October 23, 1998, report to the Security Council by UNSCOM's Executive Chairman Richard Butler questioned Iraq's claims of dismantling its WMD programs.[53] Iraq responded by halting all UNSCOM operations until Butler had been removed from his position and economic sanctions lifted. On November 5, the Security Council demanded that Iraq rescind its decision and unconditionally cooperate with inspectors.[54] To back up the demand, the United States deployed additional tactical aircraft into the Middle East and threatened air strikes. On November 14, with a U.S. strike package enroute to Iraq, Saddam succeeded in stopping the air strikes by announcing to the Russians that Iraq would fully cooperate with UNSCOM. By late November, however, Iraq had again denied access to inspectors.

In mid-December, Butler announced that Iraq had refused to cooperate with inspections, prompting UNSCOM to withdraw from Iraq altogether. The United States abandoned the pretense of coercion at this point and instead adopted a brute force strategy. On December 16, 1998, the Americans and British commenced Operation Desert Fox, a four-day restricted air campaign directed at ninety members of the Iraqi leadership, the Republican Guard, suspected nuclear and chemical facilities, and command and control targets. The crisis concluded as a foreign policy, coercion, and coercive diplomacy failure for the United States. Saddam remained in power, and the departure of UNSCOM left Iraq without monitoring or inspections until the fall of 2002.

1998: AFGHANISTAN—U.S. EMBASSY BOMBING[55]

In May 1996, Osama Bin Laden fled Sudan and relocated Al Qaeda's headquarters to Afghanistan. Encouraged by the support of the Taliban leader Omar Mullah, Bin Laden expanded Al Qaeda's network and established training camps in Afghanistan. He also issued fatwahs against the United States in August 1996 and again in February 1998. On August 7, 1998, two large truck bombs simultaneously exploded at U.S. embassies in Kenya and Tanzania, killing 220, including twelve Americans, and injuring 5,000. Al Qaeda claimed responsibility for the attacks.

Following the embassy bombings, the White House initiated a crisis by adopting a coercive strategy against the Taliban regime, demanding that Osama Bin Laden be handed over to the United States. On August 20, cruise missiles attacked three Al Qaeda compounds in Afghanistan, along with an alleged chemical plant in Sudan. The United States continued to pressure the Taliban by imposing sanctions through the Security Council. Two years of negotiations and the imposition of multilateral sanctions failed, however, to compel the Taliban to change its policies regarding Al Qaeda. U.S. coercion and coercive diplomacy failed to achieve the core foreign policy objective of convincing the Taliban to cease providing safe haven for Osama Bin Laden.

2001: AFGHANISTAN—AFGHANISTAN-USA[56]

On September 11, 2001, three hijacked airliners struck the twin World Trade Center Towers and the Pentagon, and another jet crashed in Pennsylvania enroute to Washington, D.C. The United States quickly linked the attacks to Al Qaeda, headquartered in Afghanistan under the protection of the Taliban regime.

President George W. Bush initiated a crisis by articulating the U.S. policy of pursuing terrorists and the states that sponsored them. On September 18, the Taliban's leader Mullah Omar met with the chief of Pakistan's Inter-Services Intelligence (ISI), Mahmud Ahmed, who communicated U.S. demands that Omar hand over Bin Laden and his deputies and that the Al Qaeda training camps be shut down. Bush made public these demands in a televised speech before Congress on September 20. On September 23 the Bush administration received a pessimistic field appraisal from the CIA chief at Islamabad, Pakistan, indicating that Omar was not likely to hand over Bin Laden. The United States continued preparations for military action against Al Qaeda and the Taliban, should they not fully cooperate. On September 24 the United States imposed financial sanctions against terrorist organizations.

Coercive diplomacy failed when, on October 7, the Americans commenced air strikes and deployed special operation forces to embed within the Northern Alliance and provide targeting information for U.S. air strikes. On November 9, the city of Maz-e-Sharif fell; the Taliban vacated Kabul on November 13; by December, Hamid Karzai had been named the chair of Afghanistan's interim government.

Overall, the United States achieved its core foreign policy objective of eliminating Afghanistan as a safe haven for Al Qaeda operations. Coercion and coercive diplomacy failed, however, as the White House quickly abandoned efforts to coerce the Taliban and instead adopted a brute force strategy to drive the Taliban from power. This focus on the Taliban and Afghanistan regime change was, in part, responsible for the United States missing an opportunity at Bora Bora to capture Osama Bin Laden.

2011: LIBYA—ARAB SPRING

During the Arab Spring in 2011, internal opposition in Libya led to Muammar Qaddafi ordering a crackdown against dissidents in Benghazi, as he had previously done to similar opposition in the mid-1990s. Initially France, Britain, and the United States pushed through resolutions in the Security Council imposing minimal sanctions against Libya and Qaddafi's regime. The Security Council imposed an arms embargo that UN member states immediately began to violate by arming rebels, it imposed a travel ban on only a couple of Libyan officials, and it froze the financial assets of just a few members of Qaddafi's regime and a handful of Libyan corporations.[57] Over time, however, the United States would no longer accommodate Qaddafi's actions against his own people. The Obama administration escalated the conflict into a crisis by supporting French and British efforts to garner a Security Council resolution to deter Qaddafi from attacking dissidents by authorizing a no-fly zone over Libya. To garner Russian and Chinese support, the resolution did not further authorize military force against Qaddafi's land forces. U.S. objectives were not limited, however, as President Barrack Obama identified regime change as a core objective in his March 28, 2011, address to the nation stating that Qaddafi "needed to step down from power."[58] Coercive diplomacy failed when NATO neutralized Qaddafi's military advantage by gaining air superiority and then with air strikes to disperse his conventional forces. Coercion further failed to convince Qaddafi to relinquish power, and the crisis ended only with his death and the ouster of his regime. Though coercion and coercive diplomacy failed, the United States achieved its core foreign policy objective of regime change by brute force.

APPENDIX B

ASYMMETRIC COERCION MODEL

CONSIDER TWO STATES, the challenger (C) and the target (T). An issue with a valuation of 1 is in dispute. The challenger makes demand (x) of the target, where x is continuous and ranges from 0 to 1, $(x \in [0,1])$. When $x = 0$, the outcome for the challenger and the target is $[-r_c,1]$, respectively, where r_c represents the reputation costs for the challenger. Assume both players are risk neutral, and the valuation of the demand is $v_c(x) = x$ for the challenger and $v_t(x) = 1-x$ for the target. Both states need not place the same value on the issue at hand. However, making assumptions as to the risk and valuation of the issue simplify the model without having an impact on the main outcome.

The challenger chooses demand (x) and threat (z). Threat levels range from 0 to z_{max} $(z \in [0, z_{max}])$, where $z = 0$ is no threat and z_{max} is the maximum credible threat the challenger can make, based on the relative power of the two players and the value of the issue to the challenger. The highest possible value for z_{max} is 1, where the challenger credibly adopts a brute force strategy to take the issue by force.

The challenger incurs signaling costs $s(z)$ for making offer $[x,z]$. The signaling cost function increases monotonically, is strictly convex, and is a sunk cost, whether the target concedes or resists. The challenger incurs costs to carry out threats $c_c(z)$, and the target incurs the cost of resisting $c_t(z)$, should the target reject the offer. If $z = 1$, then $c_c(z) = bf_c$, the challenger's cost of a brute force strategy, and $c_t(z) = bf_t$, the target's cost of brute force. Assume $c_c(z) = c_c \times z$, $c_t(z) = c_t \times z$, where c_c and c_t are positive coefficients. This assumption of linearity does not detract from the overall findings of the model. The target incurs reputation cost (r_t) for conceding.

The challenger's probability of coercion success $p_s(z)$, which is the likelihood the challenger will gain its objectives when the target resists, is an increasing function of threat level. This is the probability that the challenger gains demands even though the target initially resists. Assume $p_s(z) = p_s \times z$, where $p_s \in (0,1)$ and for brute force $z = 1$,

Brute force
$[p_v - bf_c, 1 - p_v - bf_t]$

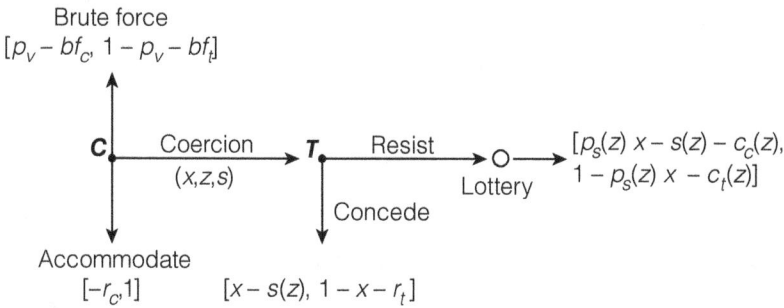

Figure B.1. Coercion model in extended form.

$p_s(1) = p_v$, where p_v is the challenger's probability of a brute force victory. Again the assumption of linearity makes calculations tractable, but it does not detract from the model's overall findings.

The game follows a sequential ultimatum protocol where the challenger moves first, making a take it or leave it offer of demands, threats, and signals (x,z,s). If $x = 0$, $z = 0$, $s = 0$, then the challenger has adopted a policy of accommodation, and if $x = 1$, $z = 1$, $s = 0$, then it has adopted a brute force strategy. If the challenger chooses a coercive offer ($x \, \varepsilon \, [0,1]$, ($z \, \varepsilon \, [0,1]$), ($s > 0$), then the target moves by either accepting or rejecting the offer. If the target resists, then the success or failure of the challenger is decided by a lottery, with the challenger obtaining its demands with $p_s(z)$ and failing with $(1 - p_s(z))$. For the lottery, the challenger's outcome $= p_s(z)[x - s(z) - c_c(z)] + (1 - p_s(z))[0 - s(z) - c_c(z)] = p_s(z) \, x - s(z) - c_c(z)$. The target's outcome $= p_s(z)[1 - x - c_t(z)] + (1 - p_s(z))[1 - c_t(z)] = 1 - p_s(z) \, x - c_t(z)$. See Figure B.1 for the model in extended form.

OPTIMIZATION OF DEMANDS AND THREATS

The optimal demand* is the highest demand the challenger makes of the target such that the target is indifferent between conceding and resisting. This optimization problem for the challenger can be rewritten in notational form in Equations 1B, 2B, and 3B:

$$1 - x^* - r_t = 1 - p_s z(x^*) - c_t z \qquad \text{Eq. 1B}$$
$$x^* = (c_t z - r_t)/(1 - p_s z) \qquad \text{Eq. 2B}$$
$$\text{Max}_z \, [(c_t z - r_t)/(1 - p_s z)] - s(z) \qquad \text{Eq. 3B}$$

Solving the optimization problem by taking first-order conditions for Eq. 3B in terms of z results in

$$c_t/(1 - p_s z^*) + p_s(c_t z^* - r_t)/(1 - p_s z^*)^2 - s_c'(z^*) = 0$$

which simplifies to

$$(c_t - p_s r_t)/(1 - p_s z^*)^2 - s_c'(z^*) = 0 \qquad \text{Eq. 4B}$$

Estimating the signaling cost function using a Taylor polynomial results in

$$s_c(z) = b_0 z^0 + b_1 z^1 + \ldots + b_n z^n = \Sigma_{i=0}^n b_i z^i \qquad \text{Eq. 5B}$$

where b_i is the coefficient for the z^ith term.

The derivative of the signaling function in terms of z is

$$s_c'(z) = b_1 + 2b_2 z + 3b_3 z^2 \ldots + nb_n z^{n-1} = \Sigma_{i=0}^n ib_i z^{i-1} \qquad \text{Eq. 6B}$$

Substituting Eq. 6B into Eq. 4B obtains the following:

$$(c_t - p_s r_t)/(1 - p_s z^*)^2 = \Sigma_{i=0}^n ib_i (z^*)^{i-1} \qquad \text{Eq. 7B}$$

Solving for z^* in general terms is not practical, however; for the case where $n = 2$, a solution set does exist. First note that the derivative of the signaling function simplifies to

$$s_c'(z) = b_1 + 2b_2 z^*$$

To keep calculations more manageable, assume $b_1 = 0$. Substituting for $s_c'(z)$ in Eq. 7B produces

$$(c_t - p_s r_t)/(1 - p_s z^*)^2 = 2b_2 z^*$$
$$(c_t - p_s r_t)/2b_2 = z^*(1 - p_s z^*)^2$$
$$(c_t - p_s r_t)/2b_2 = z^*(1 - 2p_s z^* + p_s^2 z^{*2})$$
$$z^* - 2p_s z^{*2} + p_s^2 z^{*3} - (c_t - p_s r_t)/2b_2 = 0$$
$$z^{*3} - 2p_s^{-1} z^{*2} + p_s^{-2} z^* - (c_t - p_s r_t)/2p_s^2 b_2 = 0 \qquad \text{Eq. 8B}$$

SECOND-ORDER CONDITIONS SATISFIED

To confirm this solution set is a maximum, take the second-order conditions in terms of z derived in Eq. 8B:

$$3z^{*2} - 4p_s^{-1} z^* + p_s^{-2} < 0$$
$$(3z^* - 1/p_s)(z^* - 1/p_s) < 0 \qquad \text{Eq. 9B}$$

Because $p_s \varepsilon (0,1)$ and $z^* \varepsilon [0,1]$, then $(z^* - 1/p_s) < 0$. For Eq. 10B to be true requires

$$(3z^* - 1/p_s) > 0$$
$$3z^* > 1/p_s$$
$$z^* p_s > 1/3 \qquad \text{Eq. 10B}$$

Equation 10B illustrates the inverse relationship between z^* and p_s. As the probability of coercion success increases, the maximum threat level decreases.

COERCION RANGE

The challenger's valuation of a brute force strategy where $x = 1$, $z = 1$ as depicted in Figure B.1 is

$$v_c(1,1) = v_c(\text{victory}) + v_c(\text{defeat})$$
$$v_c(1,1) = p_v(1 - bf_c) + (1 - p_v)(0 - bf_c) = p_v - p_v bf_c + p_v bf_c - bf_c$$
$$v_c(1,1) = p_v - bf_c$$

The challenger chooses coercion over brute force when

$$v_c(x^*,z^*,s(z^*)) \geq v_c(1,1)$$
$$x^* - s(z^*) \geq p_v - bf_c$$
$$x^* \geq p_v - bf_c + s(z^*) \qquad \qquad \text{Eq. 11B}$$

The target choices are concession or resistance. It chooses concession when

$$v_t(\text{concession}) \geq v_t(\text{resistance})$$
$$v_t(\text{concession}) = 1 - x^* - r_t$$
$$v_t(\text{resistance}) = 1 - p_s z(x^*) - c_t z$$

substituting

$$1 - x^* - r_t \geq 1 - p_s z(x^*) - c_t z$$
$$x^* \leq (c_t z - r_t)/(1 - p_s z) \qquad \qquad \text{Eq. 12B}$$

Note Eq. 12B is simply Eq. 2B expressed as an inequality. Combining Eq. 11B and 12B produces the coercion success range where coercion is preferred to brute force for the challenger and where concession is preferred to resistance for the target:

$$p_v - cbf_c + s(z^*) \leq x \leq (c_t z - r_t)/(1 - p_s z) \qquad \qquad \text{Eq. 13B}$$

In linear form, the coercion success range can be expressed as in Figure B.2.

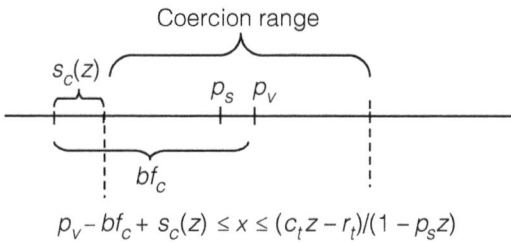

$$p_v - bf_c + s_c(z) \leq x \leq (c_t z - r_t)/(1 - p_s z)$$

Figure B.2. Coercion range.

NOTES

Chapter 1

1. Great power states identified in the Correlates of War majors2008.1.csv dataset, available at www.correlatesofwar.org: United States, Great Britain, France 1816–1940 and 1945–2008, Germany 1925–1945 and 1991–2008, Italy 1860–1943, USSR/Russia 1922–2008, China 1950–2008, and Japan 1895–1945 and 1991–2008.

2. This definition is consistent and integrates prominent theorists' definitions of coercion. Thomas Schelling defines *coercion* in terms of a punishment strategy where coercion is "the threat of damage, or of more damage to come, that can make someone yield or comply" (Thomas Schelling, *Arms and Influence*. New Haven, CT: Yale University Press, 1966, 3). Lawrence Freedman focuses on the freedom of the target of coercion to make decisions when he defines coercion as "the potential or actual application of force to influence the action of a voluntary agent" (Lawrence Freedman, *Deterrence*. Malden, MA: Polity Press, 2004, 27). Robert Pape focuses on the calculations made by the target of coercion in his definition of coercion as "efforts to change the behavior of a state by manipulating costs and benefits" (Robert Pape, *Bombing to Win*. Ithaca, NY: Cornell University Press, 1996, 4). Daniel Byman and Matthew Waxman's focus is on behavior change by the target in their definition of coercion "as the use of threatened force and at times the limited use of actual force to back up the threat, to induce an adversary to behave differently than it otherwise would" (Daniel Byman and Matthew Waxman, *The Dynamics of Coercion: American Foreign Policy and the Limits of Military Might*. Cambridge, UK: Cambridge University Press, 2002, 30). Linking the change in behavior to the challenger demands parallels to Clausewitz's definition of war, "to compel our enemy to do our will" (Carl von Clausewitz, *On War*, trans. and edited by Michael Howard and Peter Paret. Princeton, NJ: Princeton University Press, 1976, 75).

3. Seminal works on nonrational behavior include Robert Jervis, "Hypotheses on Misperception," *World Politics* 20 (1968), 454–479; Robert Jervis, Richard Ned Lebow, and Janice Gross Stein, eds., *Psychology and Deterrence* (Baltimore, MD: Johns Hopkins, 1985); Richard N. Lebow and Janice Gross Stein, *We All Lost the Cold War* (Princeton, NJ: Princeton University, 1985); James March, *A Primer on Decision Making: How Decisions Happen* (New York: Free Press, 1994); Deborah Larson, *The Origins of Containment: A Psychological Explanation* (Princeton, NJ: Princeton University Press, 1985), 24–65; Graham Allison, "Conceptual Models of the Cuban Missile Crisis," *The American Political Science Review* 63 (1969), 689–718; and Daniel Kahneman and Amos Tversky, "Prospect Theory: An Analysis of Decision under Risk," *Econometrica* 47 (1979).

4. Though rare, there are cases in World War II in which Germany coerced homeland territorial concessions from Czechoslovakia, as did the Soviet Union from Latvia, Estonia, and Lithuania.

5. See Appendix A for further details.

6. "George W. Bush: Remarks Following a Meeting with Congressional Leaders and an Exchange with Reporters, September 18, 2002." The American Presidency Project, retrieved on October 17, 2014, from www.presidency.ucsb.edu/medialist .php?presid=43.

7. Phil Haun, *On Death Ground: Why Weak States Resist Great Powers, Explaining Coercion Failure in Asymmetric Interstate Conflict.* Dissertation (Cambridge, MA; MIT, 2010), 118–120.

8. Alexander George and William Simons, *The Limits of Coercive Diplomacy* (Boulder, CO: WestView Press, 1994).

9. The Correlates of War project defines war based on the arbitrary threshold of 1,000 battlefield fatalities suffered over the course of a year. Meredith Reid Sarkees and Frank Whelon Wayman, *Resort to War: A Data Guide to Inter-State, Extra-State, Intra-state, and Non-State Wars, 1816–2007* (Washington, DC: CQ Press 2010).

Chapter 2. A Theory of Asymmetric Interstate Coercion

1. Charles Hermann, *Crises in Foreign Policy: A Simulation Analysis* (Indianapolis: Bobbs-Merrill, 1969), 414.

2. Andrew Mack, "Why Big Nations Lose Small Wars," *World Politics* 27 (1975): 181. Mack focuses on the asymmetry between an external state and a nonstate actor.

3. "Interstate System, 1816–2011 Majors 2011.csv Dataset," Correlates of War project; retrieved on December 22, 2014, from http://www.correlatesofwar.org/COW2% 20Data/SystemMembership/2011/System2011.html. Germany and Japan are included on this list as of 1990; however, they lack the nuclear and conventional military forces to threaten the survival of weak states, and so they are excluded from this list.

4. Modern great powers also have the capacity to threaten nuclear attacks that could also destroy a weak state. However, the credibility of nuclear strikes resides in its deterrent effect, and in this book the focus is primarily on the ability of the United States since World War II to compel weak states. Robert Powell, *Nuclear Deterrence Theory: The Search for Credibility* (Cambridge, UK: Cambridge University Press, 1990); and T. V. Paul, "The Nuclear Taboo and War Initiation in Regional Conflicts," *Journal of Conflict Resolution* 39 (1995): 696–717.

5. Mack, "Why Big Nations Lose Small Wars," 78. Resolve is a measure of the willingness to endure pain, suffering, or other costs; Steven Rosen, "War Power and the Willingness to Suffer," in Bruce M. Russett, ed., *Peace, War, and Numbers* (Beverly Hills, CA: Sage, 1972), 167–183.

6. Mack, "Why Big Nations Lose Small Wars," 178.

7. Since 1939, however, the idea of appeasement has been tainted by its association with Chamberlain's decision to concede Czechoslovakia to Hitler. To avoid this negative bias, this book follows Alexander George's example and instead adopts the term *accommodation*. On accommodation and appeasement, see Alexander George and William Simons, *The Limits of Coercive Diplomacy* (Boulder, CO: West View Press, 1994), 7. On appeasement, see Daniel Treisman, "Rational Appeasement," *International Organization* 58 (2004): 345; D. C. Watt, "Appeasement: The Rise of A Revisionist School?" *The Political Quarterly* 36 (1965): 191–213; and Robert Beck, "Munich's Lessons Reconsidered," *International Security* 14 (1989): 161–191.

8. "Independent Statistics and Analysis of Venezuela," U.S. Energy Information Administration; retrieved on September 26, 2012, from www.eia.gov/countries/cab .cfm?fips=VE.

9. For variant/alternative definitions of coercion, see Thomas Schelling, *Arms and Influence* (New Haven, CT: Yale University Press, 1966), 3; Lawrence Freedman, *Deterrence* (Malden, MA: Polity Press, 2004), 27; Robert Pape, *Bombing to Win* (Ithaca, NY: Cornell University Press, 1996), 4; Daniel Byman and Matthew Waxman, *The Dynamics of Coercion: American Foreign Policy and the Limits of Military Might* (Cambridge, UK: Cambridge University Press, 2002), 30; and Carl von Clausewitz, *On War*, edited by Michael Howard and Peter Paret (Princeton, NJ: Princeton University Press, 1976), 75.

10. Schelling, *Arms and Influence*, 69–78.

11. Todd Sechser, "Goliath's Curse: Coercive Threats and Asymmetric Power," *International Organization* 64 (2010): 629.

12. Glenn Snyder, *Deterrence by Denial and Punishment*, Center of International Studies Research Monograph No.1, (Princeton, NJ: Princeton University, 1958).

13. Schelling, *Arms and Influence*, 3.

14. Jonathan Shimshoni, *Israel and Conventional Deterrence: Border Warfare from 1953 to 1970* (Ithaca, NY: Cornell University Press, 1988), 6; Pape, *Bombing to Win*, 18–19.

15. "Saddam Hussein Meeting with Advisors Regarding the American Ground Attack during First Gulf War, Garnering Arab and Iraqi Support, and a Letter to Gorbachev" (Washington, DC: Conflict Research Resolution Center, National Defense University, February 24, 1991).

16. Pape, *Bombing to Win*, 15.

17. James Fearon, *Threats to Use Force: Costly Signals and Bargaining in International Crises*. Dissertation (Berkeley: University of California Berkeley, 1992), 122.

18. James Fearon, "Signaling Foreign Policy Interests: Tying Hands versus Sinking Costs," *The Journal of Conflict Resolution*, 41 (1997): 68–90.

19. See Appendix A.

20. Such deployment orders are typically kept secret. LTG John Sheehan and MG Pat Hughes, "DoD News Briefing: Lieutenant General John Sheehan," U.S. Department of Defense, Office of the Assistant Secretary of Defense (Public Affairs), October 13, 1994; and Eric Herr, "Chapter 3: Operation Vigilant Warrior," *Operation Vigilant Warrior: Conventional Deterrence Theory, Doctrine, and Practice*. Thesis (Montgomery, AL: School of Advanced Airpower Studies, 1996).

21. U.S. air operations cost approximately $20 billion in 2010 fiscal dollars; Lee Wight, "Airpower Dollars and Sense: Rethinking the Relative Costs of Combat," *Joint Forces Quarterly* 66 (2012): 55.

22. See Appendix A.

23. This definition of coercion success differs from that of Bob Pape, who codes U.S. coercion against Iraq as a success. *Bombing to Win*, 213.

24. For further discussion on efficiency, see David A. Baldwin, "The Sanctions Debate and the Logic of Choice," *International Security* 24 (1999): 80–107.

25. The United States Strategic Bombing Surveys, *Summary Report* (Montgomery, AL: Air University Press, 1987), 39.

26. Schelling, *Arms and Influence*, 3.

27. Daniel Lake, "The Limits of Coercive Airpower: NATO'S 'Victory' in Kosovo Revisited," *International Security* 34 (2009).

28. See Chapter 3.

29. This limit on force differs from that of Robert Pape, who maintains the line between a denial strategy and a brute force strategy is ambiguous, as the distinction becomes evident only ex post, with coercion succeeding if the target concedes while it still has some means to resist. Robert Pape, *Bombing to Win*, 15.

30. George, *The Limits of Coercive Diplomacy*, 10–11. In addition, for more on coercive diplomacy, see Alexander George, *Forceful Persuasion: Coercive Diplomacy as*

an Alternative to War (Washington, DC: U.S. Institute of Peace Press, 1991); Robert J. Art and Patrick M. Cronin, *The United States and Coercive Diplomacy* (Washington, DC: United States Institute of Peace Press, 2003); Daniel Byman and Matthew Waxman, *The Dynamics of Coercion: American Foreign Policy and the Limits of Military Might* (New York: Cambridge University Press, 2002); Jack S. Levy, "Deterrence and Coercive Diplomacy: The Contributions of Alexander George," *Political Psychology* 29:4 (2008); Peter Viggo Jakobsen, "Coercive Diplomacy," *Contemporary Security Studies* (Oxford, UK: Oxford University Press, 2010); Kenneth A Schultz, *Democracy and Coercive Diplomacy* (New York: Cambridge University Press, 2001); and Bruce Jentleson "Coercive Diplomacy: Scope and Limits in the Contemporary World," *The Stanley Foundation Policy Analysis Brief*, December 2006, retrieved on October 1, 2012 from http://stanleyfdn.org/publications/pab/pab06CoerDip.pdf.

31. George, *The Limits of Coercive Diplomacy*, 2.

32. Ibid., 17.

33. For more details on North Korean nuclear crisis, see Appendix A.

34. Gary Clyde Hufbauer, Jeffrey J. Schott, Kimberly Ann Elliott, and Barbara Oegg, *Economic Sanctions Reconsidered*, 3rd ed. (Washington, DC: Institute for International Economics, 2007); Robert Pape, "Why Economic Sanctions Do Not Work," *International Security* 22 (1997): 90–136; Kimberly Elliot, "The Sanctions Glass: Half Full or Completely Empty," *International Security* 22 (1998): 50–65; and Daniel Drezner, "Hidden Hand of Economic Coercion," *International Organization* 57 (2003): 643–659.

35. Peterson Institute for International Economics, "Case Studies in Sanctions and Terrorism: Case 99-1 US and UN v. Afghanistan (Taliban)"; retrieved on October 2, 2012, from http://www.piie.com/research/topics/sanctions/afghanistan.cfm.

36. Previous research in coercion and related subfields assumes either the level of demands or threats to be fixed and examines the effect on outcome by varying the remaining variable. Pape, *Bombing to Win*, holds demands fixed in his coercion model by only evaluating those cases where important/territorial issues are at stake. James D. Fearon, "Rationalist Explanations for War," *International Organization* 49 (1995), holds constant for the threat level only considering the threat of war and probability of victory being fixed. Another example of holding threat fixed is Suzanne Werner, "Deterring Intervention: The Stakes of War and Third-Party Involvement," *American Journal of Political Science* 44 (2000): 720–732. A final game theoretic example of holding demands fixed while varying military force is Branislav L. Slantchev, "Military Coercion in Interstate Crises," *American Political Science Review* 99 (2005): 533–547.

37. U.S. State Department, *Rambouillet Agreement: Interim Agreement for Peace and Self-Government in Kosovo*; available at www.state.gov/www/regions/eur/ksvo_rambouillet_text.html.

38. Kenneth Waltz, *Theory of International Politics* (Boston: McGraw-Hill, 1979).

39. Seminal works on nonrational behavior include Robert Jervis, "Hypotheses on Misperception," *World Politics* 20 (1968), 454–479; Robert Jervis, Richard Ned Lebow, and Janice Gross Stein, eds., *Psychology and Deterrence* (Baltimore: Johns Hopkins, 1985); Richard N. Lebow and Janice Gross Stein, *We All Lost the Cold War* (Princeton, NJ: Princeton University, 1985); James March, *A Primer on Decision Making: How Decisions Happen* (New York: Free Press, 1994); Deborah Larson, *The Origins of Containment: A Psychological Explanation* (Princeton, NJ: Princeton University Press, 1985), 24–65; Graham Allison, "Conceptual Models of the Cuban Missile Crisis," *The American Political Science Review* 63 (1969): 689–718; and Daniel Kahneman and Amos Tversky, "Prospect Theory: An Analysis of Decision Under Risk," *Econometrica* 47 (1979).

40. Steven R. David, "Explaining Third World Alignment," *World Politics* 43 (1991): 233–256.

41. Assuming the target initially controls the issue restricts demands to being compellent. This restriction is not that limiting because all the asymmetric interstate crises for the United States since World War II have been cases of compellence.

42. For a critique on actual reputation costs, see Daryl G. Press, *Calculating Credibility: How Leaders Assess Military Threats* (Ithaca, NY: Cornell University Press, 2005), 3.

43. James Morrow, "The Strategic Setting of Choices: Signaling, Commitment and Negotation," in *Strategic Choice and International Relations* (Princeton, NJ: Princeton University Press, 1999), 78–114.

44. James F. Schnabel, *U.S. Army in the Korean War: Policy and Direction: The First Year* (Washington DC: Center for Military History, U.S. Army, 1992).

45. Patricia L. Sullivan, "War Aims and War Outcomes: Why Powerful States Lose Limited Wars," *Journal of Conflict Resolution* 51 (2007): 496–524.

46. The model does not allow for partial concession by the target, though in the real world this clearly happens. For instance, following the El Dorado Canyon air raid in April 1986 against Libya, Qaddafi partially met U.S. demands by no longer overtly supporting terrorist groups, although still conducting clandestine operations.

47. Additional costs to the target are the losses incurred from the challenger's signaling. This is not addressed for two reasons. First, for symbolic signals the targets costs are negligible. The second reason is that for signals generated by limited force the costs to the target are incurred prior to the target's decision making and therefore are not part of the target's calculations. These additional costs would matter only if the limited use of force destroyed a significant portion of the issue at stake.

48. Judith Miller, "In Rare Talks with Libyans, U.S. Airs Views on Sanctions," *New York Times*, June 12, 1999.

49. 2010 dollars; see Wight, "Airpower Dollars and Sense," 55.

50. A common modeling assumption is made in that, if indifferent between concession or resistance, the target will concede.

51. The lower bounded solution represents accommodation.

52. The coercion range is closely related to James Fearon's concept of a bargaining range. Fearon, "Rationalist Explanations for War," 387.

53. Schelling, *Arms and Influence*, 1.

54. Art, *The United States and Coercive Diplomacy*, 375.

55. Adam Meirowitz and Anne Sartori, "Strategic Uncertainty as a Cause of War," *Quarterly Journal of Political Science* (2008); Robert Powell, "Bargaining and Learning while Fighting," *American Journal of Political Science* 48 (2004): 344–361; and Fearon, "Rationalist Explanations for War," 391.

56. Fearon, "Rationalist Explanations for War," 381.

57. Ibid., 398–399.

58. Geoffrey Blainey, *The Causes of War*, 3rd ed. (New York: The Free Press, 1988), 56.

59. James Fearon, "Bargaining, Enforcement, and International Cooperation," *International Organization* 52 (1998): 269–305; Robert Axelrod and Robert Keohane, "Achieving Cooperation under Anarchy: Strategies and Institutions," *World Politics* 38 (1985): 226–254; and Kenneth Oye, "Explaining Cooperation under Anarchy: Hypotheses and Strategies," *World Politics* 38 (1985): 1–24.

60. Robert Powell, "War as a Commitment Problem," *International Organization* 60 (2006): 169–203; Robert Powell, "The Inefficient Use of Power: Costly Conflict with Complete Information," *American Political Science Review* 98 (2004): 231–241; and Fearon, "Rationalist Explanations for War," 381.

61. Todd Sechser, *Winning without a Fight: Power, Reputation, and Compellent Threats in International Crises*. Dissertation (Palo Alto, CA: Stanford University, 2007), 5.

62. Fearon, "Rationalist Explanations for War," 379.

63. Ibid., 382.

64. Powell, "War as a Commitment Problem," 169–203.

65. Ron Hassner, "To Halve and to Hold: Conflicts over Sacred Space and the Problem of Indivisibility," *Security Studies* 12 (2003): 4; Monica Toft, "Issue Indivisibility and Time Horizons as Rationalist Explanations for War," *Security Studies* 15 (2006): 38; and Stacie Goddard, "Uncommon Ground: Indivisible Territory and the Politics of Legitimacy," *International Organization* 60 (2006): 35–68.

66. Ron Hassner, *War on Sacred Grounds* (Ithaca, NY: Cornell, 2009).

Chapter 3

1. George W. Bush, "Address to the Nation, 17 March 2003."

2. Bob Woodward, *Plan of Attack* (New York: Simon and Schuster, 2002), 343.

3. Robert Pape, *Bombing to Win* (Ithaca, NY: Cornell University Press, 1996), 21.

4. Kenneth Waltz, *Theory of International Politics* (Boston: McGraw-Hill, 1979), 95–96.

5. Tanisha M. Fazal, *State Death: The Politics and Geography of Conquest, Occupation, and Annexation* (Princeton, NJ: Princeton University Press, 2007).

6. Fazal, *State Death*, 23.

7. Challenger and target resolve affect a brute force war outcome, and in the asymmetric war model these are factored into the costs of fighting the war.

8. Andrew Mack, "Why Big Nations Lose Small Wars: The Politics of Asymmetric Conflict," *World Politics* 27 (1975): 175–200.

9. Waltz, *Theory of International Politics*, 95–96.

10. Stephen D. Krasner, "Sharing Sovereignty: New Institutions for Collapsed and Failing States," *International Security* 29 (2004): 85–120.

11. Waltz, *Theory of International Politics*, 95–96.

12. Sun Tzu, *Art of War*, translated by Samuel B. Griffith (New York: Oxford University Press, 1971), 68.

13. See Chapter 5.

14. Thomas Schelling, *Arms and Influence* (New Haven, CT: Yale University Press, 1966), 132.

15. Krasner, "Sharing Sovereignty," 70.

16. Stephen D. Krasner, "Sovereignty," *Foreign Policy*, 122 (2001): 28.

17. For more discussion on Warsaw Pact states see Fazal, *State Death*, 24.

18. A possible exception is North Korea's decision to withdraw from the Non-Proliferation Treaty in 2003.

19. Krasner, "Sharing Sovereignty," 70.

20. Ibid., 115.

21. Fazal, *State Death*, 23.

22. Zanzibar had its autonomy for only four tumultuous months following its independence from Britain in December 1963. Its minority Arab government was quickly overthrown by an ethnic African majority who selected Abeid Karume as their president. Violence and ethnic killings ensued; lacking the security forces to restore order, Karume sought military and police assistance from neighboring Tanganyika. With this assistance, he was able to gain control of the islands in return for which he agreed to share sovereignty with Tanganyika, thus forming the state of Tanzania in April 1964; see Ian Speller, "An African Cuba? Britain and the Zanzibar Revolution," *Journal of Imperial and Commonwealth History* 35 (2007): 19. This exceptional case of

a tiny island chain with a population of less than one-quarter million bargaining away its autonomy illustrates how rarely modern states agree to give up their international sovereignty.

23. See Chapter 5.

24. The asymmetric coercion model also assumes that the powerful challenger acts as a unitary actor. Here the focus is on target concessions, and so the unitary actor assumption is relaxed only on the target.

25. For a discussion on political survival, see Bruce Bueno de Mesquita, Alastair Smith, Randolph M. Siverson, and James D. Morrow, *The Logic of Political Survival* (Cambridge MA: MIT Press, 2003), chapter 1.

26. Steven David, "Explaining Third World Alignment," *World Politics* 43 (1991): 233–256.

27. The logic of domestic costs of a coup is similar to that of audience costs, see James Fearon, "Domestic Political Audiences and the Escalation of International Disputes," *American Political Science Review* 88 (1994): 577–592; and James D. Fearon, "Rationalist Explanations for War," *International Organization* 49 (1995): 379–414.

28. James D. Fearon, "Domestic Politics, Foreign Policy, and Theories of International Relations," *Annual Review of Political Science* 1 (1998): 299.

29. Barbara Geddes, *Paradigms and Sand Castles* (Lansing, MI: University of Michigan Press, 2003); and Barbara Geddes, "Authoritarian Breakdown: Empirical Test of a Game Theoretic Argument." Paper presented at the Annual Meeting of the American Political Science Association, Atlanta, GA, April 1999.

30. Geddes, "Authoritarian breakdown"; and Jessica L. Weeks, "Autocratic Audience Costs: Regime Type and Signaling Resolve," *International Organization* 62 (2008): 35–64.

31. Phil Haun, *On Death Ground: Why Weak States Resist Great Powers Explaining Coercion Failure in Asymmetric Interstate Conflict*. Dissertation (Cambridge, MA: MIT, 2010), 133.

32. Barry Posen provided this insight.

33. The challenger does also sacrifice any advantage it may have attained from the element of surprise.

34. Saddam attempted to concede to the UN resolutions, but not to U.S. demands that Iraq withdraw from Kuwait in forty-eight hours and thus abandon its heavy weapons.

35. George Bush and Brent Scowcroft, *A World Transformed* (New York: Alfred A. Knopf, 1998), 437.

36. For more on the initiation of crises to justify war see Richard Ned Lebow, *Between Peace and War* (Baltimore: Johns Hopkins University Press, 1981), chapter 2.

Chapter 4

1. UN Security Council, "Resolution 678 (1990), 29 November 1990"; George W. Bush and Brent Scowcroft, *A World Transformed* (New York: Alfred A. Knopf, 1998), 394; and Bob Woodward, *The Commanders* (New York: Simon and Schuster, 1991), 320.

2. UN Security Council, "Security Council Holds Iraq in 'Material Breach' of Disarmament Obligations, Offers Final Chance to Comply, Unanimously Adopting Resolution 1441 (2002)."

3. Phoebe Marr, *The Modern History of Iraq* (Boulder, CO: Westview Publishing, 2004), 204; Effraim Karsh and Inari Rautsi, *Saddam Hussein: A Political Biography* (New York: The Free Press, 1991), 202; and "Iraq Banking on Credit," *The Economist*, September 30, 1989.

4. "Aziz Assails Kuwait, UAE in Letter to Kblil," *Baghdad Domestic Service* JN1807103390, July 18, 1990. Saddam later claimed the loans had always been intended as free aid. George L. Piro, "Interview with Saddam Hussein: Casual Conversation, 24 February 2004," *Saddam Hussein Talks to the FBI*, National Security Archives Electronic Briefing Book No. 279.

5. Conflict Records Research Center, "Meetings between Saddam Hussein and the Yemeni President, and between Saddam and Dr. George Habash, Secretary General of the Popular Front of the Palestinian Liberation" (Washington, DC: National Defense University, August–September 1990), SH_MISC_D_000_652.

6. "Assembly Issues Statement on Kuwait, UAE," *Baghdad Domestic Service* FBIS-NES-90-140, July 20, 1990; and Youssef Ibrahim, "OPEC Meets Today: Talks Are Clouded by Iraq's Threat to Kuwait," *New York Times*, July 25, 1990.

7. "Confrontation in the Gulf; Excerpts from Iraqi Document on Meeting with U.S. Envoy," *New York Times*, September 23, 1990; and James Baker, *The Politics of Diplomacy* (New York: G. P. Putnam's Sons, 1995), 272.

8. Organization of the Petroleum Exporting Companies, "World Proven Oil Reserves by Country 1980–2004"; retrieved on July 28, 2013, from www.opec.org/library/annual%20statistical%20bulletin/interactive/2004/filez/XL/T33.HTM.

9. Energy Information Administration, "What Drives Crude Oil Prices"; retrieved on July 28, 2013, from www.eia.gov/finance/markets/reports_presentations/eia_what_drives_crude_oil_prices.pdf.

10. George Bush, "A Collective Effort to Reverse Iraqi Aggression," *U.S. Department of State Dispatch*, September 10, 1990; and "Test of Saddam Husayn Initiative on Situation," *Baghdad Domestic Service* FBIS-NES 90-156, August 13, 1990.

11. George Bush, "Remarks at the Aspen Institute Symposium in Aspen Colorado," August 2, 1990 (Dallas, TX: George W. Bush Presidential Library and Museum).

12. George Bush, "Address before a Joint Session of the Congress on the Persian Gulf Crisis and Budget Deficit," September 11, 1990 (Dallas, TX: George W. Bush Presidential Library and Museum).

13. George Bush, "Address to the Nation Announcing the Deployment of United States Armed Forces to Saudi Arabia," August 8, 1990. (Dallas, TX: George W. Bush Presidential Library and Museum).

14. Bush, *A World Transformed*, 322; and Baker, *The Politics of Diplomacy*, 277–278.

15. UN Security Council, "Resolution 660 (1990), 2 August 1990."

16. Clyde H. Farnsworth, "The Iraqi Invasion; Bush, in Freezing Assets, Bars $30 Billion to Hussein," *New York Times*, August 3, 1990; and UN Security Council, "Resolution 661 (1990), 6 August 1990."

17. George Bush, "Address to the Nation Announcing the Deployment of United States Armed Forces to Saudi Arabia," August 8, 1990 (Dallas, TX: George W. Bush Presidential Library and Museum).

18. Michael R. Gordon and Bernard E. Trainor, *The Generals' War* (Boston: Little, Brown, 1995), 76.

19. Thomas A. Keaney and Eliot A. Cohen, *Gulf War Air Power Survey Summary Report* (Washington, DC: U.S. Government Printing Services, 1993), 36.

20. John Warden, "The Enemy as a System," *Airpower Journal*, Spring 1995.

21. Keaney, *Gulf War Air Power Survey Summary Report*, 4.

22. Conflict Records Research Center, "Meeting between Saddam Hussein and the Soviet Delegation" (Washington, DC: National Defense University, October 10, 1990), SH-PDWN-D-000-533.

23. "Text of Saddam Husayn Initiative on Situation," *Baghdad Domestic Service*, FBIS-NES-90-156, August 13, 1990.

24. "Foreigners to Receive Food under Ration System," *Baghdad Domestic Service*, FBIS-NES-90-191, October 2, 1990.

25. George Lardner, "CIA Director Sanctions Need 9 More Months; Webster Says Embargo Offers No Guarantee," *Washington Post*, December 6, 1990; and Peter Passell, "Confrontation in the Gulf; How Vulnerable Is Iraq?" *New York Times*, August 20, 1990.

26. Conflict Records Research Center, "Meeting between Saddam Hussein and the Soviet Delegation" (Washington, DC: National Defense University, October 10, 1990), SH-PDWN-D-000-533.

27. *Baghdad Domestic Service*, "RCC, Ba'th Party Issue Statement on 'Battle,'" FBIS-NES-90-184, September 21, 1990.

28. Yevgeny Primakov, *Russian Crossroads: Toward the New Millennium* (New Haven, CT: Yale University Press, 2004), 48–49.

29. Ibid., 49; the Primakov quote is also supported by audio tapes made of the meeting. Conflict Records Research Center, "Meeting between Saddam Hussein and the Soviet Delegation" (Washington, DC: National Defense University, October 10, 1990), SH-PDWN-D-000-533.

30. Bush, *A World Transformed*, 377.

31. Woodward, *The Commanders*, 319.

32. An early start to the air war was imperative to prevent the ground war from taking place during the Kuwaiti summer. The extreme heat of Middle Eastern summer conditions would make combat operations dangerous and difficult for coalition forces, particularly those deploying from Western Europe.

33. Woodward, *The Commanders*, 322.

34. "Mideast Tensions; Excepts from Gulf Testimony" *New York Times*, 4 December 1990.

35. George Bush, "President's News Conference on the Persian Gulf Crisis" November 8, 1990 (Dallas, TX: George W. Bush Presidential Library and Museum).

36. Michael R. Gordon, "Mideast Tensions; Nunn, Citing 'Rush' to War, Assails Decision to Drop Troop Rotation Plan," *New York Times*, November 12, 1990.

37. Maureen Dowd, "Mideast Tensions; Americans More Wary of Gulf Policy, Poll Finds," *New York Times*, November 20, 1990.

38. Defense Secretary Dick Cheney, General Powell, and Secretary of State James Baker all testified before congressional committees in December 1991. Baker, *The Politics of Diplomacy*, 340.

39. The twelve UNSC resolutions pertaining to Iraq and Kuwait are summarized as follows: UNSC Resolution 660, August 2, 1990, demanded that "Iraq withdraw immediately and unconditionally all its forces to the positions in which they were located on 1 August 1990"; UNSC Resolution 661, August 6, 1990, "Decides that all States shall prevent . . . the import into their territories of all commodities and products originating in Iraq or Kuwait"; UNSC Resolution 662, August 9, 1990, "Decides that annexation of Kuwait by Iraq under any form and whatever pretext has no legal validity"; UNSC Resolution 664, August 18, 1990, "Demands that Iraq permit and facilitate the immediate departure from Kuwait and Iraq of third-State nationals"; UNSC Resolution 665, August 25, 1990, "Calls upon those Member States co-operating with the Government of Kuwait which are deploying maritime forces to the area to use such measures . . . to halt all inward and outward maritime shipping"; Resolution 666, September 13, 1990, "Decides that if the Committee, after receiving the reports from the Secretary-General, determines that circumstances have arisen in which there is an urgent humanitarian need to supply foodstuffs to Iraq and Kuwait in order to relieve human suffering, it will report promptly to the Council its decision as to how such need should be met"; Resolution 667, September 16, 1990, "Further demands that Iraq immediately protect the safety and well-being of diplomatic and consular personnel

and premises in Kuwait and in Iraq"; UNSC Resolution 669, September 24, 1990, introduced a letter from the president of the council to the secretary-general of a special report pertaining to Jordan, which requested "relief under Article 50 of the Charter of the United Nations from the effects resulting from implementation of the measures required under Security Council resolution 661"; UNSC Resolution 670, September 25, 1990, "Decides that all States . . . shall deny permission to any aircraft to take off from their territory if the aircraft would carry any cargo to or from Iraq or Kuwait other than food in humanitarian circumstances"; UNSC Resolution 674, October 29, 1990, "Demands that the Iraqi authorities and occupying forces immediately cease and desist from taking third-State nationals hostage, mistreating and oppressing Kuwaiti and third-State nationals and any other actions . . . relative to the Geneva Convention relative to the Protection of Civilian Person in Time of War"; UNSC Resolution 677, November 28, 1990, "Gravely concerned at the ongoing attempt by Iraq to alter the demographic composition of Kuwait . . . mandates the Secretary-General to take custody of a copy of the population register of Kuwait"; and UNSC Resolution 678, November 29, 1990, "Demands that Iraq comply fully with resolution 660 (1990) and all subsequent relevant resolutions, and decides, while maintaining all its decisions, to allow Iraq one final opportunity, as a pause of good will, to do so." It also "Authorizes Member States co-operating with the Government of Kuwait, unless Iraq on or before 15 January 1991 fully implements . . . the above-mentioned resolutions, to use all necessary means to uphold and implement resolution 660 and all subsequent relevant resolutions and to restore international peace and security in the area."

40. UN Security Council, "Resolution 678 (1990), 29 November 1990"; Bush, *A World Transformed*, 394; and Woodward, *The Commanders*, 320.

41. *Baghdad Domestic Service*, "Saddam Interview with ABC," FBIS-NES-90-223, November 19, 1990; and Bush, *A World Transformed*, 420.

42. Baker, *The Politics of Diplomacy*, 361.

43. Adam Clymer, "Confrontation in the Gulf; Congress Acts to Authorize War in Gulf; Margins are 5 votes in Senate, 67 in House," *New York Times*, January 13, 1991.

44. Baghdad Domestic Service, "Saddam Husayn Calls up 250,000 More Troops," FBIS-NES-90-223, November 19, 1990. U.S. intelligence estimated 540,000 Iraqi troops in the Kuwait Area of Operation, although postwar analysis estimates the number at 336,000. Eliot A. Cohen and Thomas A. Keaney, *Gulf War Airpower Survey*, vol. 1 (Washington, DC: U.S. Government Printing Office, 1993), 233.

45. James Lemoyne, "Mideast Tensions: Saudis Say Jerusalem Killings Could Weaken Alliance against Iraq," *New York Times*, October 10, 1990; Paul Lewis, "Mideast Tensions; U.S. Joins in 2d Vote at UN to Criticize Israel over 21 Slain," *New York Times*, October 25, 1990; and R. W. Apple, "Two-Front Campaign; For Mr. Bush, Holding Together Fragile Coalitions Is Getting Harder," *New York Times*, October 14 1990.

46. Primakov, *Russian Crossroads*, 55.

47. Bush, *A World Transformed*, 383; and George H. W. Bush, "Responding to Iraqi Aggression in the Gulf," *National Security Directive 54*, 2, January 15, 1991.

48. Baghdad INA, "Briefs Saddam on U.S. 'Intransigence,'" FBIS-NES-91-008, January 10, 1991; Baghdad INA, "U.S. 'Solely Responsible' for Failure," FBIS-NES-91-008, January 11, 1991; and Baghdad Domestic Service, "Saddam Visits Battlefront, Speaks to Troops," FBIS-NES-91-011, January 15, 1991.

49. Baghdad Domestic Service, "Saddam Relieves Shanshal of Defense Minister Post," FBIS-NES-90-239; and Amman Domestic Service December 12, 1990, "Saddam Says Tel Aviv 'First Target' in War," FBIS-NES-90-247, December 24, 1990. Saddam also mentioned to the Soviet Union in October 1990 his intention to strike Israel. Conflict Records Research Center, "Iraq Officials Discussing the Retreat of Iraq Troops from Kuwait and the Greediness of the United States of America" (Washington, DC: National Defense University, late 1991) SH-SHTP-A-000-669.

50. Saddam would later claim that the attacks on Israel were not to draw it into the conflict and split the coalition but rather that hurting Israel would get the United States to stop the war. Piro, "Interview with Saddam Hussein: Casual Conversation, March 3, 2004."

51. Baghdad Domestic Service, "Saddam Interview with German TV," FBIS-NES-90-247, December 24, 1990; and Baghdad Domestic Service, "Saddam Interview with Spanish TV," December 27, 1990.

52. Baghdad Domestic Service, "Saddam Says Jihad Has Put Nation on Right Paths," FBIS-NES-91-001, January 1, 1991; Baghdad Domestic Service, "Saddam Speaks," FBIS-NES-91-004, January 6, 1991; Baghdad Domestic Service, "Saddam Stresses 'High' State of Readiness," FBIS-NES-91-005, January 8, 1991; and Baghdad Domestic Service, "Saddam Addresses Reception," FBIS-NES-91-007, January 10, 1991.

53. Baghdad Domestic Service, "Saddam: U.S. to Swim 'in Blood' if War Starts," FBIS-NES-91-007, January 10, 1991.

54. Kevin Woods, James Lacey, and Williamson Murray, "Saddam's Delusions," *Foreign Affairs* 85 (2006): 2.

55. General Powell makes a similar assessment of Saddam's priorities. Colin Powell, *My American Journey* (New York: Random House, 1995), 490.

56. Max Weber, *The Theory of Social and Economic Organization* (London: Collier-Macmillan, 1947), 358, 360.

57. Piro, "Interview with Saddam Hussein: Casual Conversation, February 24, 2004."

58. Bush, *A World Transformed*, 323.

59. Keaney, *Gulf War Airpower Survey*, vol. 2, 115.

60. Keaney, *Gulf War Airpower Survey Summary*, 59.

61. Keaney, *Gulf War Airpower Survey Summary*, 61; and Keaney, *Gulf War Airpower Survey*, vol. 2, 176.

62. Keaney, *Gulf War Airpower Survey*, vol. 5, 642–643.

63. Keaney, *Gulf War Airpower Survey*, vol. 2, 387.

64. John Horner and Tom Clancy, *Every Man a Tiger: The Gulf War Air Campaign* (New York: Berkeley Publishing, 1999), 372–374.

65. Baghdad Domestic Service, "Commander Cables Saddam," FBIS-NES-91-013, January 18, 1991; and Keaney, *Gulf War Airpower Survey Summary*, 84.

66. Notes taken by an unknown senior Iraqi military officer commanding the overall Scud missile operations. Conflict Records Research Center, "Statements about the Iraq War in 1991" (Washington, DC: National Defense University, Unknown), SH-MISC-D-000-298.

67. Bush, *A World Transformed*, 451–457; Baker, *The Politics of Diplomacy*, 385–390; and Kevin M. Woods, David D. Palkki, and Mark E. Stout, *The Saddam Tapes* (Cambridge, UK: Cambridge University Press, 2011), 183–188.

68. Woods, *The Saddam Tapes*, 195.

69. Baghdad Domestic Service, "Gorbachev Letter Proposes Withdrawal from Kuwait," FBIS-NES-91-014, January 22, 1991; and Baghdad INA, "Text of Saddam's Reply," FBIS-NES-91-014, January 22, 1991.

70. At the commencement of the war the United States estimated Iraq strength at 540,000 troops, 4000 tanks, and 3000 artillery pieces. However, a combination of undermanning and Iraqi troops disserting or on leave significantly reduced the Iraqi troop strength. Keaney, *Gulf War Airpower Survey*, vol. 2, 254; and Keaney, *Gulf War Airpower Survey Summary*, 106.

71. Keaney, *Gulf War Airpower Survey*, vol. 2, 269–270.

72. Ibid., 269–286, 331. There were 1,460 sorties dedicated to the Scud hunt, compared to the 19,073 strike sorties in the KTO.

73. George N. Lewis, Steve Fetter, and Lisbeth Gronlund, "Casualties and Damage from Scud Attacks in the 1991 Gulf War," Defense and Arms Control Studies Working Paper, MIT Center for International Studies, March 1993, 43.

74. Joel Brinkley, "War in the Gulf: Israel; No Immediate Retaliation Israelis Say," *New York Times*, January 24, 1991; Andrew Rosenthal, "War in the Gulf: The Overview; Pentagon Is Confident on War but Says Iraqis Remain Potent; Sees No Imminent Land Attack," *New York Times*, January 24, 1991.

75. Piro, "Interview with Saddam Hussein: Casual Conversation, March 3, 2004."

76. Baghdad INA, "INA Reports Saddam Visit to Southern Front," FBIS-NES-91-016, January 24, 1991; and Gordon, *The Generals' War*, 269.

77. These comments were made by Saddam on what might have been accomplished militarily had Al Khafji taken place earlier and to a larger degree. Conflict Records Research Center, "Iraq Officials Discussing the Retreat of Iraq Troops from

Kuwait and the Greediness of the United States of America" (Washington, DC: National Defense University, late 1991) SH-SHTP-A-000-669; and Kevin M. Woods, *The Mother of All Battles: Saddam Hussein's Strategic Plan for the Persian Gulf War* (Annapolis, MD: Naval Institute Press, 2008), 289.

78. Air Force Studies and Analyses Agency, *Airpower and the Iraqi Offensive at Khafji* CDROM (Washington, DC: AFSAA Force Application Division, 1997).

79. Keaney, *Gulf War Airpower Survey*, vol. 2, 275, 280.

80. Ibid., 275, 208–209, 280.

81. Horner, *Every Man a Tiger*, 389.

82. Keaney, *Gulf War Airpower Survey*, vol. 2, 68.

83. Tunis Domestic Service, "Holds News Conference," FBIS-NES-91-029, February 11, 1991.

84. Cairo Domestic Service, "Soviet Envoy Arrives in Baghdad 11 Feb," FBIS-NES-91-029, February 12, 1991.

85. Baghdad Domestic Service, "Saddam Spells Out Position," FBIS-NES-91-030, February 13, 1991; and Primakov, *Russian Crossroads*, 68.

86. Bush, *A World Transformed*, 471.

87. Keaney, *Gulf War Airpower Survey*, vol. 2, 281.

88. Norman Schwarzkopf and Peter Petre, *It Doesn't Take a Hero* (New York: Bantom, 1992), 506.

89. Mikhail Gorbachev, *Memoirs* (New York: Doubleday, 1995), 560.

90. Gorbachev, *Memoirs*, 561; and "War in the Gulf: Moscow's Statement; Transcript of Comments on Soviet Peace Proposal," *New York Times*, February 23, 1991.

91. Primakov, *Russian Crossroads*, 70.

92. Gorbachev, *Memoirs*, 562.

93. Primakov, *Russian Crossroads*, 70; Gorbachev, *Memoirs*, 562; Serge Schmemann, "War in the Gulf: Diplomacy; Soviets Say Iraq Accepts Kuwait Pullout Linked to Truce and an End to Sanctions; Bush Rejects Conditions: War Is to Go On," *New York Times*, February 22, 1991; and "War in the Gulf: Soviet Statement; Moscow's Statement on the Iraqis' Response," *New York Times*, February 22, 1991.

94. Piro, "Interview with Saddam Hussein: Casual Conversation, February 24, 2004."

95. Woods, *The Mother of All Battles*, 269.

96. "War in the Gulf: U.S. Statement; Transcript of White House Statement and News Conference on Soviet Plan," *New York Times*, February 22, 1991.

97. Bush, *A World Transformed*, 474–476; and Maureen Dowd, "War in the Gulf: White House; Pressing Demands," *New York Times*, February 22, 1991.

98. George H. W. Bush, "Remarks on the Persian Gulf Conflict 1991-02-22," February 22, 1991 (Dallas, TX: George Bush Presidential Library and Museum).

99. George H. W. Bush, "NSD 54: Responding to Iraqi Aggression in the Gulf, 15 January 1991."

100. International Institute for Strategic Studies, *The Military Balance 1990–1991*, 105.

101. Ibid., 107.

102. The 100,000 additional troops were estimated by assessing the additional heavy weapons that might have been saved by the Iraqi Army without a hasty retreat.

103. Keaney, *Gulf War Air Power Survey*, vol. 5, 27–32.

104. Baghdad Domestic Service, "RCC Statement on Bush's 'Disgraceful' Ultimatum," FBIS-NES-91-037, February 25, 1991; Associated Press, "War in the Gulf: Iraq; U.S. Peace Terms Denounced by Iraq," *New York Times*, February 23, 1991; and "War in the Gulf; Statement by Iraqi Revolutionary Council," *New York Times*, February 23, 1991.

105. Marlin Fitzwater, "Statement by Press Secretary Fitzwater on the Persian Gulf Conflict," February 23, 1991 (Dallas, TX: George W. Bush Presidential Library and Museum.

106. Gordon, *The Generals' War*, 355–358.

107. Associated Press, "War in the Gulf: The Marines; 2 Divisions Said to Near Kuwait City," *New York Times*, February 25, 1991; and Gordon, *The Generals' War*, 369.

108. Baghdad Domestic Service, "Official Spokesman Says 'Withdrawal Order' Given," FBIS-NES-91-038, February 26, 1991; and Patrick Tyler, "War in the Gulf: The Overview; Iraq Orders Troops to Leave Kuwait but U.S. Pursues Battlefield Gains," *New York Times*, February 26, 1991.

109. Keaney, *Gulf War Airpower Survey*, vol. 2, 261. The Republican Guard divisions that were deployed in reserve along the Iraq–Kuwait border did not suffer as high attrition, losing approximately half of their tanks, APCs, and artillery. Anthony H. Cordesman, *Iraq and the War of Sanctions: Conventional Threats and Weapons of Mass Destruction* (Westport, CT: Praeger, 1991), 68.

110. On the lead-up to the war and during the war the estimates of troops and equipment was higher than these figures, which were adjusted downward following after action assessments of Iraqi military capabilities. Gordon, *The Generals' War*, 459; International Institute for Strategic Studies, *The Military Balance, 1991–1992*; and International Institute for Strategic Studies, *The Military Balance, 1990–1991*.

111. Gorbachev, *Memoirs*, 564.

112. Marlin Fitzwater, "Statement by Press Secretary Fitzwater on the Persian Gulf Conflict," February 25, 1991 (Dallas, TX: George W. Bush Presidential Library and Museum).

113. "War in the Gulf: Diplomacy; Texts of Iraqi Letters," *New York Times*, February 28, 1991.

114. R. W. Apple, "After the War: The Overview; U.S. Says Iraqi Generals Agree to Demands 'on All Matters'; Early P.O.W. Release Expected," *New York Times*, March 4, 1991; Gordon, *The Generals' War*, 444–447; and Schwarzkopf, *It Doesn't Take a Hero*, 557–568.

115. Geoffrey Blainey, *The Causes of War*, 3rd ed. (New York: Free Press, 1988), 56.

116. Jessica Weeks, "Autocratic Audience Costs: Regime Type and Signaling Resolve," *International Organization* 62 (2008): 35–64.

117. Steven David, "Explaining Third World Alignment," *World Politics* 43 (1991): 233–256.

118. Kevin M. Woods and Mark E. Stout, "Saddam's Perceptions and Misperceptions: The Case of 'Desert Storm,'" *Journal of Strategic Studies* 33 (February 2010): 12–13.

119. "UN Security Council Resolution 706," August 15, 1991; David Malone, *The International Struggle over Iraq: Politics in the U.N. Security Council 1980–2005* (New York: Oxford University Press, 2006), 117; "UN Security Council Resolution 986," April 14, 1995; UN Security Council, "Letter dated 20 May 1996 from the Secretary-General Addressed to the President of the Security Council"; United Nations, "Office of the Iraq Programme Oil-for-Food," May 20, 1996; "UN Security Council Resolution 1153," February 20, 1998; "UN Security Council Resolution 1284," December 17, 1999; and U.S. Department of State, "Fact Sheet: Iraq-Goods Review List," January 14, 2003.

120. Michael Knights, *Cradle of Conflict: Iraq and the Birth of the Modern U.S. Military* (Annapolis, MD: Naval Institute Press, 2005), 200–210.

121. For an overview of military operations during this period see Knight, *Cradle of Conflict*, 150–250.

122. U.S. Congress, "H.R. 4655: Iraq Liberation Act of 1998," October 31, 1998; Edward Walker, "Origins of the Iraq Regime Change Policy," Department of State, January 23, 2001.

123. Donald Rumsfeld, "U.S. Department of Defense, Notes from Donald Rumsfeld [Iraq War Planning]," The National Security Archives, *The Iraq War Ten Years After*, Document 3.

124. George W. Bush, "State of the Union Address," January 29, 2002; George W. Bush, "West Point Graduation Speech," June 1, 2002; and The White House, *The National Security Strategy of the United States of America*, September 2002.

125. "After the Attack: Reaction from Around the World," *New York Times*, September 13, 2001.

126. Bush, "State of the Union Address," January 29, 2002.

127. Bush, "President Bush Delivers Remarks at West Point," June 1, 2002.

128. Bush, "Address to the United Nations," September 12, 2002.

129. U.S. Congress, "H.J. Res. 114: Authorization for Use of Military Force against Iraq Resolution of 2002," October 16, 2002.

130. UN Press Release SC/7564, "Security Council Holds Iraq in 'Material Breach' of Disarmament Obligations, Offers Final Chance to Comply, Unanimously Adopting Resolution 1441 (2002)," November 8, 2002.

131. "UN Security Council Resolution 1441," November 8, 2002; and Bob Woodward, *Plan of Attack* (New York: Simon & Schuster, 2004), 223–227.

132. George W. Bush, "State of the Union Address," January 29, 2003.

133. David Sander and Thom Shanker, "Threats and Responses: The Military; War Plan for Iraq Calls for Big Force and Quick Strikes," *New York Times*, November 10, 2002; and Tommy Franks, *American Soldier* (New York: Harper-Collins, 2004), 409–410.

134. Iraqi Survey Group, "Realizing Saddam's Veiled WMD Intent," *Final Report*, October 2004.

135. Iraqi Foreign Minister Naji Sabri, "Letter to Secretary General Annan," September 19, 2002.

136. Baghdad Babil, "Editorial by Abd-al-Razzaq al-Dulaymi Reacting to Iraq's Decision to Comply with UN Resolution 1441," FBIS-NES-2002-1114, November 14, 2002.

137. Central Intelligence Agency, "Misreading Intentions: Iraq's Reaction to Inspections Created Picture of Deception," January 5, 2006; and The National Security Archives, *The Iraq War Ten Years After*, Document 12, January 5, 2005.

138. Iraqi Survey Group, "Realizing Saddam's Veiled WMD Intent," 2004.

139. Kevin Woods, Michael Pease, Mark Stout, Williamson Murray, and James G. Lacy, *Iraqi Perspectives Project: A View of Operation Iraqi Freedom from Saddam's Senior Leadership* (Annapolis, MD: Naval Institute Press, 2006), 92.

140. International Institute for Strategic Studies, *The Military Balance 2001–2002*.

141. Woods, *Iraqi Perspectives Project*, 93.

142. "Iraqi Minister of Defense Calls for an Investigation into Why Documents of a WMD Nature Were Found by a UN Inspection Team," SH-GMID-D-000-890, July 1998.

143. George Piro, "Comprehensive Report of the Special Advisor to the DCI on Iraq's WMD," with Addendums (Duelfer Report), April 2005 [Excerpt]; and The National Security Archives, *The Iraq War Ten Years After*, Document 11, April 2005.

144. Central Intelligence Agency, "Misreading Intentions," Document 12.

145. Saddam stated in an interview to FBI agent Piro that Iran was a major threat to Iraq, but Israel was only a threat to the entire Arab World and not specifically to Iraq. Piro, "Interview with Saddam Hussein: Casual Conversation, June 11, 2004."

146. Dan Collins, "Saddam's Secret Weapon? Fedayeen, a Tough Militia Force Personally Loyal to Iraqi Leader," *CBS News*, March 24, 2003.

147. Woods, *Iraqi Perspectives Project*, 26.

148. Barbara Geddes, "Authoritarian Breakdown: Empirical Test of a Game Theoretic Argument." Paper presented at the Annual Meeting of the American Political Science Association, Atlanta, GA, April 1999.

149. Woods, *Iraqi Perspectives Project*, 92.

150. Barton Gelman, "Keeping the U.S. First; Pentagon Would Preclude a Rival Superpower," *Washington Post*, March 11, 1992, A1; and Paul Wolfowitz, Donald Rumsfeld, Richard Armitage, et al., "Open Letter to President William J. Clinton," *Project for the New American Century*, January 26, 1998.

151. Bob Woodward, *Bush at War* (New York: Simon and Schuster, 2002), 60.

152. George W. Bush, "Remarks Following a Meeting with Congressional Leaders and an Exchange with Reporters," *The American Presidency Project*, September 18, 2002.

153. Ari Fleischer, "White House Press Conference," *CNN*, December 2, 2002.

154. "Eyes on Iraq; In Cheney's Words: The Administration Case for Removing Saddam Hussein," *New York Times*, August 27, 2002.

155. "Threats and Responses: Report by Iraq; Iraq Arms Report Has Big Omissions, U.S. Officials Say," *New York Times*, December 13, 2002.

156. Hans Blix and Mohamed ElBaradei, "News Update on Iraq Inspections" IAEA, U.N. Chiefs Brief Press, December 19, 2002.

157. Colin Powell, "Threats and Responses; in Powell's Words: 'We Are Disappointed, but We Are Not Deceived' by Iraq," *New York Times*, December 20, 2002.

158. Central Intelligence Agency, "Misreading Intentions," Document 12, January 5, 2006.

159. United Kingdom, "Matthew Rycroft, Private Secretary to the Prime Minister, Cabinet Minutes of Discussion, S 195/02, July 23, 2002"; The National Security Archives, *The Iraq War Ten Years After*, Document 5, August 15, 2002; and Woodward, *Plan of Attack*, 418, 424.

160. Hans Blix, "An Update on Inspection," UN Security Council, January 27, 2003.

161. Bush, "State of the Union Address," January 28, 2003.

162. Powell, "U.N. Security Council Presentation," February 5, 2003.

163. "Threats and Responses: Washington; U.S. Ready to Back New UN Measure on Iraq, Bush Says," *New York Times*, February 7, 2003.

164. Piro, "Interview with Saddam Hussein: Casual Conversation, June 11, 2004."

165. Iraqi Survey Group, "Realizing Saddam's Veiled WMD Intent," *Final Report*, October 2004.

166. Ian Fisher, "State of the Union: Baghdad; Iraqi Aide Pledges 'Extra Effort' to Cooperate with Inspectors," *New York Times*, January 29, 2003.

167. Iraqi Survey Group, *Final Report*, 63.

168. Piro, "Interview with Saddam Hussein: Casual Conversation, February 13, 2004"; and Neil MacFarquhar, "Threats and Responses: Iraq; Sandbags Already on Streets. Baghdad Is a City in Waiting," *New York Times*, March 12, 2003.

169. "Threats and Responses: Diplomacy; 3 Members of NATO and Russia Resist U.S. on Iraq Plans," *New York Times*, February 11, 2003.

170. Todd Purdham, *A Time of Our Choosing* (New York: Time Books, 2003), 74.

171. George W. Bush, "President George Bush Discusses Iraq in National Press Conference," March 6, 2003.

172. "U.N. diplomats told to be ready for possible vote Tuesday," *CNN World*, March 7, 2003.

173. "Text: Azores Summit Statement," *BBC News*, March 16, 2003.

174. George W. Bush, "Bush: 'Leave Iraq within 48 hours,'" *CNN World*, March 17, 2003.

175. Woodward, *Plan of Attack*, 180.

176. Eric Schmitt and James Dao, "Airpower Alone Can't Defeat Iraq, Rumsfeld Asserts," *New York Times*, July 31, 2002; and Woodward, *Plan of Attack*, 71, 81, 113.

177. Woodward, *Plan of Attack*, 343.

178. Michael Gordon, "Threats and Responses: Military Plans; Allies Will Move in, Even if Saddam Hussein Moves Out," *New York Times*, March 18, 2003.

179. Woodward, *Plan of Attack*, 314; and John Burns, "Threats and Responses: Amman; Jordan Pressing U.S. to Offer Exile to Hussein and His Aides if They Yield Power in Iraq," *New York Times*, February 12, 2003.

180. Piro, "Interview with Saddam Hussein: Casual Conversation, June 28, 2004."

181. John Burns, "Threats and Responses: Amman; Jordan Pressing U.S. to Offer Exile to Hussein and His Aides if They Yield Power in Iraq," *New York Times*, February 12, 2003; and John Burns, "Threats and Responses: Iraq; Defiant Response," *New York Times*, March 20, 2003.

182. Franks, *American Soldier*, 348.

183. Anthony Cordesman, *The Iraq War: Strategy, Tactics, and Military Lessons* (Washington, DC: Center for Strategic and International Studies, 2004), 40, 46.

184. Dan Collins, "Saddam's Secret Weapon? Fedayeen, a Tough Militia Force Personally Loyal to Iraqi Leader," *CBS News*, March 24, 2003.

185. Cordesman, *The Iraq War*, 47.

186. Michael R. Gordon and Bernard E. Trainor, *Cobra II* (New York: Random House, 2006), 124.

187. Woodward, *Plan of Attack*, 30.

188. Franks, *American Soldier*, 349; and Gordon and Trainor, *Cobra II*, 28.

189. Franks, *American Soldier*, 371.

190. David Sander and Thom Shanker, "Threats and Responses: The Military; War Plan for Iraq Calls for Big Force and Quick Strikes," *New York Times*, November 10, 2002; and Franks, *American Soldier*, 409–410.

191. Cordesman, *The Iraq War*, 36–37.

192. Ibid., 24.

193. Dexter Filkins, "Threats and Responses: Ankara; Turkish Deputies Refuse to Accept American Troops," *New York Times*, March 2, 2003.

194. Gordon and Trainor, *Cobra II*, 69.

195. George W. Bush, "President Bush Addresses the Nation," March 19, 2003.

196. "Timeline: Iraq," *BBC*, May 21, 2011.

197. George W. Bush, "Bush Makes Historic Speech aboard Warship," *CNN*, May 1, 2003.

198. Elizabeth Bumiller "White House Cuts Estimate of Cost of War with Iraq," *New York Times*, January 2, 2003; Tim Russert, "Interview with Vice-President Dick Cheney," *Meet the Press*, March 16, 2003; Eric Schmitt, "Pentagon Contradicts General on Iraq Occupation Force's Size," *New York Times*, February 28, 2003; and *Congressional Record*, "Proceedings and Debates of the 108th Congress, Second Session Volume 150-Part 14," September 17, 2004, 18661.

199. Amy Belasco, *The Cost of Iraq, Afghanistan, and Other Global War on Terror Operations since 9/11* (Washington, DC: Congressional Research Service, March 29, 2011), 25; and Brown University's Costs of War Project, "Iraq War: 190,000 lives, $2.2 trillion," March 14, 2013. This estimate includes all casualties as well as the economic costs to the United States for the war.

200. Mark Thompson, "March Was First Month without U.S. Fatalities in Iraq or Afghanistan in 11 Years," *Time*, April 1, 2014; retrieved on December 22, 2014, from http://time.com/45160/zero-us-fatalities-iraq-afghanistan-11-years/.

201. Eric Schmitt, "Threats and Responses." *New York Times*, November 6, 2002.

202. Iraqi Coalition Provisional Authority, "Order Number 1: De-Ba'athification of Iraqi Society, May 16, 2003," and "Order Number 2: Dissolution of Entities, August 23, 2003," The National Security Archives, *The Iraq War Ten Years After*, Document 5, August 15, 2002.

Chapter 5

1. Lenard Cohen, *Serpent in the Bosom: The Rise and Fall of Slobodan Milosevic* (Boulder, CO: Westview, 2001), 31.

2. Adam LeBor, *Milosevic: A Biography* (New Haven, CT: Yale University Press, 2004), 71.

3. Reuters, "Protest Staged by Serbs in an Albanian Region," *New York Times*, April 26, 1987.

4. Cohen, *Serpent in the Bosom*, 77.

5. Marlise Simons, "Upheaval in the East: Yugoslavia; Yugoslav Communists Vote to End Party's Monopoly," *New York Times*, January 23, 1990.

6. Robert Hayden, "Constitutional Nationalism in the Formerly Yugoslav Republics," *Slavic Review* 51 (1992): 660.

7. Tim Judah and Anne McElvoy, "Belgrade Ready for Border Sacrifices to Preserve Unity," *The Times, London*, July 16, 1991.

8. Chuck Sudetic, "2 Yugoslav States Vote Independence to Press Demands," *New York Times*, June 26, 1991. Macedonia declared its independence on September 8, 1991; Susan Woodward, *Balkan Tragedy: Chaos and Dissolution after the Cold War* (Washington, D.C.: Brookings, 1995), 119.

9. UN Security Council, "Resolution 713 (1991), 25 September 1991."

10. Laura Silber and Allan Little, *Yugoslavia: Death of A Nation* (New York: Penguin, 1995), 177.

11. Alan Cowell, "Fighting Slows under Yugoslav Truce; Hopes for a Peace Force," *New York Times*, November 25, 1991; and Cohen, *Serpent in the Bosom*, 204–205.

12. Steven Burg and Paul Shoup, *The War in Bosnia-Herzegovinia: Ethnic Conflict and International Intervention* (New York: M. E. Sharpe, 1999), 26, 45–48.

13. David Binder, "Serbia and Croatia Agree to Another Cease-Fire; 4th Independence Move," *New York Times*, October 16, 1991.

14. Chuck Sudetic, "Serbs Proclaim Autonomy in Another Yugoslav Region," *New York Times*, January 10, 1992.

15. Burg, *The War in Bosnia-Herzegovinia*, 106–107.

16. Chuck Sudetic, "Turnout in Bosnia Signals Independence," *New York Times*, March 2, 1992.

17. The European Community became the European Union in 1993.

18. George Bush, "Statement on United States Recognition of the Former Yugoslav Republics," April 7, 1992 (Dallas, TX: George W. Bush Presidential Library and Museum).

19. UN Security Council, "Resolution 757 (1992), 30 May 1992."

20. Silber, *Yugoslavia: Death of A Nation*, 251; and Chuck Sudetic, "Breaking Cease-Fire, Serbs Launch Attacks into Bosnia," *New York Times*, April 15, 1992.

21. UN High Commission on Refugees, "Chapter 7: Internally Displaced Persons, Lessons from Bosnia and Herzegovina," *The State of the World's Refugees 2006*; retrieved on December 22, 2014, from www.unhcr.org/4a4dc1a89.html.

22. Burg, *The War in Bosnia-Herzegovinia*, 199; UN Security Council, "Resolution 764 (1992), 13 July 1992"; UN Security Council, "Resolution 770 (1992), 13 August 1992"; and UN Security Council, "Resolution 776 (1992), 14 September 1992."

23. B. G. Ramcharan, editor, *The International Conference on the Former Yugoslavia: Official Papers Volume 1* (Rjinland in Leiden, Netherlands: Kluwer Law International, 1997), 29–57.

24. F. Watson, "Peace Proposals for Bosnia-Herzegovina," *House of Commons Library*, London Research Paper No. 93/35, March 23, 1993.

25. Burg, *The War in Bosnia-Herzegovinia*, 224, 242.

26. "Serbia Per Capita GDP," *United Nations Data*; retrieved on January 8, 2013, from http://data.un.org/.

27. "Tragedy Continues with 'No Sign of Abatement,'" *UN Chronicle* 30 (1993): 10–20.

28. Gregory Hall, "The Politics of Autocracy: Serbia under Slobodan Milosevic," *East European Quarterly* 33 (1999): 240–243.

29. John Burns, "Serbs Reported Willing to Allow Muslims to Leave Overrun Area," *New York Times*, March 5, 1993.

30. UN Security Council, "Resolution 820 (1993), 17 April 1993"; and Burg, *The War in Bosnia-Herzegovinia*, 244.

31. Burg, *The War in Bosnia-Herzegovinia*, 246.

32. John Burns, "Bosnian Serbs' Leaders Meet to Ratify Vote Rejecting Peace Plan," *New York Times*, May 20, 1993.

33. David Binder, "Pariah as Patriot; Ratko Mladic," *New York Times Magazine*, September 4, 1994.

34. James Baker, *The Politics of Diplomacy* (New York: G. P. Putnam's Sons, 1995), 636, 648–649.

35. Gwen Ifill, "Conflict in the Balkans; Clinton Takes Aggressive Stances on Role of U.S. in Bosnia Conflict," *New York Times*, August 10, 1992.

36. Burg, *The War in Bosnia-Herzegovinia*, 269–286.

37. Roger Cohen, "Terror in Sarajevo; NATO to Hold Emergency Talks on Sarajevo Attack," *New York Times*, February 7, 1994.

38. North Atlantic Council, "Decisions Taken at the Meeting of the North Atlantic Council," Press Release, February 9, 1994, 15.

39. Michael Gordon, "Conflict in the Balkans; NATO Craft Down 4 Serb Warplanes Attacking Bosnia," *New York Times*, March 1, 1994.

40. William Schmidt, "Croats and Muslims Reach Truce to End the Other Bosnia Conflict," *New York Times*, February 24, 1994.

41. The UN Security Council declared Sarajevo, Bihac, Tuzla, Gorazde, Zepa, and Srebrenica as safe areas. UN Security Council, "Resolution 824 (1993), 6 May 1993."

42. Chuck Sudetic, "Conflict in the Balkans; The Overview; U.S. Planes Bomb Serbian Position for a Second Day," *New York Times*, April 12, 1994; Chuck Sudetic, "Conflict in the Balkans; The Offensive; Serbs Down a British Jet over Gorazde," *New York Times*, April 17, 1994; and Chuck Sudetic, "Serbian Soldiers Seize Guns Held by UN, Then Return Most," *New York Times*, April 20, 1994.

43. Steven Greenhouse, "U.S., Britain and Russia Form Group to Press Bosnia Accord," *New York Times*, April 26, 1994.

44. Steven Greenhouse, "Peace Outline Has Its Flaws, Bosnians Say," *New York Times*, May 15, 1994.

45. Burg, *The War in Bosnia-Herzegovinia*, 300.

46. Alan Riding, "Bosnian Serbs Said to Reject Mediators' Partition Plan," *New York Times*, July 21, 1994.

47. Richard Lyons, "UN Security Council Weighs Rewarding or Punishing Serbs," *New York Times*, August 5, 1994; Silber, *Yugoslavia: Death of A Nation*, 341–343; and Stephen Engelberg and Eric Schmitt, "Conflict in the Balkans: The Serbian Role; Western Officials Say Serbia Helps Bosnian Comrades," *New York Times*, June 11, 1995.

48. Manojla Milovanovic, former Republica Srpska General, interview by Phil Haun, May 12, 2010, Banja Luka, Bosnia.

49. Filip Svarm, "The Silence of the General," *Vreme News digest Agency*, 152, August 22, 1994.

50. Silber, *Yugoslavia: Death of A Nation*, 131.

51. Michael Gordon, "President Orders End to Enforcing Bosnian Embargo," *New York Times*, November 11, 1994; and Cees Wiebes, *Intelligence and the War in Bosnia 1992–1995* (Berlin: Lit Verlag, 2003).

52. Michael Gordon, "U.S. Proposes Exclusion Zone in Bosnia Town," *New York Times*, November 18, 1994.

53. UN Security Council, "Resolution 959 (1994), November 19, 1994"; and Burg, *The War in Bosnia-Herzegovinia*, 300.

54. Roger Cohen, "Fighting Rages as NATO Debates How to Protect Bosnian Enclave," *New York Times*, November 25, 1994.

55. Roger Cohen, "France Seeking Plan for Ending Bosnia Mission," *New York Times*, December 8, 1994.

56. Warren Bass, "The Triage of Dayton," *Foreign Affairs* 77 (1998): 99–100.

57. Roger Cohen, "Seeking Carter Visit, Bosnia Serbs Ease Up," *New York Times*, December 17, 1994; Tony Barber, "Bosnian Serbs 'Succeeded in Outwitting Jimmy Carter,'" *The Independent*, December 22, 1994; and The Carter Center, "President Carter Helps Restart Peace Efforts in Bosnia-Herzegovina," September 1, 1994.

58. Roger Cohen, "Croatia Hits Area Rebel Serbs Hold, Crossing UN Lines," *New York Times*, May 2, 1995; Roger Cohen, "Croats Attack, Serbs Flee and Another Town Is Uprooted," *New York Times*, May 5, 1995; and Roger Cohen, "April 3–May 6: A New Phase; The Balkan Wars Heat Up as Croatia Takes the Field to Roll Back Serbs," *New York Times*, May 7, 1995.

59. Roger Cohen, "U.S. Cooling Ties to Croatia after Winking at Its Buildup," *New York Times*, October 28, 1995; William Clinton, *My Life* (New York: Knopf, 2004), 667; and Wiebes, *Intelligence and the War in Bosnia 1992–1995*.

60. Burg, *The War in Bosnia-Herzegovinia*, 328.

61. Roger Cohen, "NATO May Be Called on to Silence Guns in Sarajevo," *New York Times*, May 25, 1995.

62. Roger Cohen, "NATO Jets Bomb Arms Depot at Bosnian Serb Headquarters," *New York Times*, May 26, 1995; and Alison Mitchell, "Clinton Defends NATO Air Strikes in Bosnia and Calls on Serbs to Free UN Hostages," *New York Times*, May 27, 1995.

63. Roger Cohen, "Peacekeeping vs. an Intractable War," *New York Times*, June 11, 1995.

64. John Darnton, "Clinton's Offer of Troops Pleases Europe," *New York Times*, June 2, 1995; Eric Schmitt, "Briton Suggests UN May Leave Bosnia," *New York Times*, June 9, 1995; and Warren Christopher, *In the Stream of History: Shaping Foreign Policy for a New Era* (Stanford, CA: Stanford University Press, 1998), 348.

65. Stephen Kinzer, "U.S.–Serb Talks Suspended," *New York Times*, June 8, 1995.

66. Chris Hedges, "Bosnian Serbs Overrun Town Protected by UN," *New York Times*, July 12, 1995; Chris Hedges, "Second 'Safe Area' in Eastern Bosnia Overrun by Serbs," *New York Times*, July 20, 1995; and International Committee of the Red Cross, "Bosnia-Herzegovina: Ten Years after Fall of Srebrenica, Families of Missing Persons Continue to Suffer," July 8, 2005.

67. Bob Woodward, *The Choice* (New York: Simon & Schuster, 1996), 258, 263.

68. Todd Purdam, "Clinton, Facing Objections, Refines Narrow Conditions for Using Troops in Bosnia," *New York Times*, June 4, 1995; and Woodward, *The Choice*, 258.

69. Woodward, *The Choice*, 269; Elaine Sciolino, "House, Like Senate, Votes to Halt Bosnia Embargo," *New York Times*, August 2, 1995; Stephen Engelberg, "How Events Drew U.S. into Balkans," *New York Times*, August 19, 1995; and Richard Holbrooke, *To End a War* (New York: Modern, 1998), 68.

70. John Darnton, "Allies Warn Bosnian Serbs of 'Substantial' Air Strikes if UN Enclave Is Attacked: Accord in London," *New York Times*, July 22, 1995.

71. Craig Whitney, "Allied Extending Shield to Protect All Bosnia Havens," *New York Times*, August 2, 1995.

72. Barbara Crossette, "UN Military Aides Given Right to Approve Attacks," *New York Times*, July 27, 1995.

73. The U.S. plan was a seven-point initiative calling for (1) a comprehensive peace settlement; (2) three-way recognition of Bosnia, Croatia, and the Federal Republic of Yugoslavia (Serbia and Montenegro); (3) lifting of economic sanctions on the Federal Republic of Yugoslavia; (4) peaceful return of eastern Slavonia to Croatia; (5) cease-fire and end of all offensive operations; (6) reaffirmation of Contact Group plan for 51/49 split of territory to Bosnian Federation (Muslim and Croats) and Bosnian Serbs, respectively; (7) comprehensive economic program for regional reconstruction. Holbrooke, *To End a War*, 74.

74. It also included the withdrawal of U.S. support for the Muslim–Croat Bosnian Federation if they also refused the plan, Woodward, *The Choice*, 268–269.

75. Burg, *The War in Bosnia-Herzegovinia*, 344–345.

76. Steven Greenhouse, "U.S. Criticizes Croatia, but Only Halfheartedly, for Attack on Serbs," *New York Times*, August 5, 1995; and Raymond Bonner, "Croatia Declares Victory in Rebel Area," *New York Times*, August 7, 1995.

77. Jane Perlez, "Serb Chief's Response to Events Is Restrained," *New York Times*, August 6, 1995; Jane Perlez, "Croatian Serbs Blame Belgrade for Their Rout," *New York Times*, August 11, 1995; and Silber, *Yugoslavia: Death of a Nation*, 357.

78. Mike O'Connor, "Bosnian Army Presses Offensive against Rebel Serbs in Central Region," *New York Times*, August 14, 1995; and Karl Mueller, "The Demise of Yugoslavia and the Destruction of Bosnia: Strategic Causes, Effects, and Responses." In *Deliberate Force: A Case Study in Effective Air Campaigning*, edited by Robert Owen (Maxwell AFB, AL: Air University Press, 2000), 27.

79. Milovanovic interview, May 12, 2010; and Jane Perlez, "Bosnian Serb Leader Demotes Commander," *New York Times*, August 6, 1995.

80. Milovanovic interview, May 12, 2010; Jane Perlez, "Power Struggle of 2 Top Leaders Grows," *New York Times*, August 7, 1995; and Jane Perlez, "Bosnian Serbs, Angry at Setback and Tired of War, Blame Leaders," *New York Times*, August 14, 1995.

81. Holbrooke, *To End a War*, 4.

82. Christopher Bennett, *Yugoslavia's Bloody Collapse* (New York: NYU Press, 1995), 133.

83. Barbara Geddes, "Authoritarian Breakdown: Empirical Test of a Game Theoretic Argument." Paper presented at the Annual Meeting of the American Political Science Association, Atlanta, GA, April 1999.

84. Silber, *Yugoslavia: Death of A Nation*, 365; David Dittmer and Stephen Dawkins, *Deliberate Force: NATO'S First Extended Air Operation* (Washington, DC: Center for Naval Analysis, 1998), 19; Steven Greenhouse, "U.S. Officials Say Bosnian Serbs Face NATO Attack if Talks Stall," *New York Times*, August 28, 1995; Roger Cohen, "Shelling Kills Dozens in Sarajevo; U.S. Urges NATO to Strike Serbs," *New York Times*, August 29, 1995; and Mueller, "The Demise of Yugoslavia," 27.

85. Derek Chollet, *The Road to the Dayton Accords* (New York: Palgrave, 2005), 63; and Silber, *Yugoslavia: Death of A Nation*, 365.

86. Milovanovic interview, May 12, 2010.

87. Silber, *Yugoslavia: Death of a Nation*, 366.

88. Alan Cowell, "Fighting Slows under Yugoslav Truce; Hopes for a Peace Force," *New York Times*, November 25, 1991.

89. Robert Cohen, "Shelling Kills Dozens in Sarajevo; U.S. Urges NATO to Strike Serbs," *New York Times*, August 29, 1995; Richard Sargent, "Deliberate Force Targeting." In *Deliberate Force: A Case Study in Effective Air Campaigning*, edited by Robert

Owen, 285; Dittmer, *Deliberate Force: NATO'S First Extended Air Operation*, 10–11; and AFSOUTH Fact Sheets, *Operation Deliberate Force*, December 16, 2002.

90. Sargent, "Deliberate Force Targeting," 337; Dittmer, *Deliberate Force: NATO'S First Extended Air Operation*, 28; and Roger Cohen, "NATO Jets Attack Serbian Positions around Sarajevo," *New York Times*, August 30, 1995.

91. Milovanovic interview, May 10, 2010; Dittmer, *Deliberate Force*, 22; and David Nichols, "Bosnia: UN and NATO," *Royal United Services Institute Journal* 141 (1996): 35–36.

92. Dittmer, *Deliberate Force*, 23.

93. Roger Cohen, "Serbs Balk but NATO Delays Raids," *New York Times*, September 3, 1995.

94. Yugoslav Daily Survey, "Mladic Calls for Urgent Meeting of Bosnia Factions," and "Former U.S. President Announces Serb Compliance with NATO Demands," September 4 and 5, 1995; and Roger Cohen, "A NATO Deadline in Bosnia Passes without Attack," *New York Times*, September 5, 1995.

95. Admiral Leighton Smith, "Transcript of Press Conference Admiral Leighton W. Smith, Commander in Chief, Allied Forces Southern Europe: NATO Recommences Air Strikes against Bosnian Serbs," September 6, 1995.

96. "Division within Unity," *New York Times*, September 9, 1995.

97. Roger Cohen, "Croatia Expands Its Power in Bosnia," *New York Times*, September 16, 1995.

98. Dittmer, *Deliberate Force*, 37.

99. Trevor Murray, "Transcript Deliberate Force Press Brief, 11 September 1995"; and Mike O'Connor, "Bosnian Serb Civilians Flee Joint Muslim-Croat Attack," *New York Times*, September 14, 1995.

100. Holbrooke, *To End a War*, 152; and Elaine Sciolino, "Sarajevo Pact: Diplomacy on a Roll," *New York Times*, September 15, 1995.

101. Burg, *The War in Bosnia-Herzegovinia*, 354.

102. Dittmer, *Deliberate Force*, 41–45.

103. Chris Hedges, "Extent of Croat-Bosnia Advance Threatens U.S.-Brokered Peace," *New York Times*, September 19, 1995.

104. Holbrooke, *To End a War*, 160.

105. Stephen Kinzer, "Bosnian Serbs Fend off Croatian and Muslim Attacks," *New York Times*, September 22, 1995; Holbrooke, *To End A War*, 164; and Milovanovic interview, May 10, 2010.

106. Chris Hedges, "Negotiator Says Cease-Fire in Bosnia Is Unlikely Soon," *New York Times*, October 4, 1995.

107. William Clinton, "Remarks Announcing Agreement on a Cease-Fire in Bosnia-Herzegovina and an Exchange with Reporters," *The American Presidency Project*, October 5, 1995; retrieved on December 22, 2014, from www.presidency.ucsb.edu/.

108. Roger Cohen, "Balkan Leaders Face an Hour for Painful Choices," *New York Times*, November 1, 1995; Elaine Sciolino, "Accord Reached to End the War in Bosnia," *New York Times*, November 22, 1995; Office of the High Representative and EU Special Representative, "The General Framework Agreement for Peace in Bosnia and Herzegovina," *Dayton Peace Agreement*, December 14, 1995; and Craig Whitney, "Balkan Foes Sign Peace Pact, Dividing an Unpacified Bosnia," *New York Times*, December 15, 1995.

109. Holbrooke, *To End a War*, 238.

110. "Bosnian Serbs Moderate Confrontation," *New York Times*, March 9, 1999.

111. Holbrooke, *To End a War*, 310.

112. Mike O'Connor, "Bosnian Army Presses Offensive against Rebel Serbs in Central Region," *New York Times*, August 14, 1995; and Mueller, "The Demise of Yugoslavia," 27.

113. Jane Perlez, "Bosnian Serb Leader Demotes Commander," *New York Times*, August 6, 1995.

114. Holbrooke, *To End a War*, 299.

115. Dittmer, *Deliberate Force: NATO'S First Extended Air Operation*, 41–45.

116. Cohen, *Serpent in the Bosom*, 204–205.

117. Ibid., 255.

118. Rexhep Selimi, interview by Phil Haun, May 18, 2010, Pristina, Kosovo. Rexhep was among twenty founding members of KLA. He was the first KLA member along with two others to come forward publicly on November 28, 1997. He served as the operations officer (G-3) at KLA headquarters in 1998 and as inspector general for the KLA during the NATO air campaign. Cohen, *Serpent in the Bosom*, 234.

119. Tim Judah, *Kosovo: War and Revenge* (New Haven, CT: Yale University Press, 2000), 128–129.

120. International Crisis Group, *Kosovo Spring: Europe Report*, March 1998; retrieved on December 22, 2014, from www.crisisgroup.org/en/regions/europe/balkans/kosovo/032-kosovo-spring.aspx.

121. Fred Abrahams and Elizabeth Anderson, *Humanitarian Law Violations in Kosovo* (New York: Human Rights Watch, 1998), 28.

122. Rexhep Selimi interview, May 18, 2010.

123. Nebi Qena, interview by Phil Haun, May 17, 2010, Pristina, Kosovo. Nebi is a Kosovar Albanian and Associated Press reporter who covered the Kosovo campaign in Pristina until April 3, 1999, when he and his family were forced to leave the city. He continued reporting from Macedonia until the end of the war in June 1999.

124. The Contact Group consisted of the United States, Russia, Great Britain, France, Germany, and Italy. Steven Erlanger, "Albright Tours Europe to Whip Up Resolve to Punish Serbia," *New York Times*, March 8, 1998; and Steven Erlanger, "Sanctions on Yugoslavia," *New York Times*, March 10, 1998.

125. Philip Shenon, "U.S. Dispatches Its Balkans Mediator with a Warning," *New York Times*, May 9, 1998.

126. Jim Lehrer, *The News Hour Transcript*, June 5, 1998; and Chris Hedges, "Slim Hope for U.S. Peace Effort as Milosevic Flouts Accord," *New York Times*, June 1, 1998.

127. Chris Hedges, "Refugees from Kosovo Cite a Bitter Choice: Flee or Die," *New York Times*, June 5, 1998.

128. Craig Whitney, "Offensive by Serbia Puts Allies in War Room," *New York Times*, June 11, 1998; Marc Weller, *The Crisis in Kosovo 1989–1999; From the Dissolution of Yugoslavia to Rambouillet and the Outbreak of Hostilities*, vol. 1 (Cambridge, UK: Book Systems Plus, 1999), 236.

129. Rexhep Selimi interview, May 18, 2010.

130. Mike O'Connor, "Kosovo Refugees: Pawns in a NATO-Serb Clash?" *New York Times*, August 24, 1998.

131. OSCE Kosovo Verification Mission, *Kosovo as Seen as Told: An Analysis of the Human Rights Findings of the OSCE Kosovo Verification Mission October 1998 to June 1999* (Warsaw: Office for Democratic Institutions and Human Rights, 1999), chapters 1, 4; and "Ibrahim Rugova: Pacifist at the Crossroads," *BBC Online Network*, May 5, 1999.

132. Judah, *Kosovo*, 154.

133. Rexhep Selimi interview, May 18 2010; and Judah, *Kosovo*, 66–67.

134. Agim Ceku, interview by Phil Haun, May 17, 2010, Pristina, Kosovo. General Lieutenant Agim Ceku, the commanding officer of the KLA, claimed that the KLA made a mistake by not meeting with Holbrooke in the summer of 1998. Rexhep Selimi, a founding member of the KLA, agreed with Ceku but pointed out that, in the summer of 1998, although the KLA had reorganized militarily, it did not have a political wing capable of engaging the U.S. diplomatically. Rexhep Selimi interview, May 18, 2010.

135. Rexhep Selimi interview, May 18, 2010.

136. Ibid.

137. UN Security Council, "Resolution 1199 (1998), September 23, 1998."

138. Strobe Talbott, *The Russia Hand* (New York: Random House, 2002), 302.

139. Agim Ceku interview, May 17, 2010; Jeffrey Smith, "Turnaround in Kosovo: Rebels Bounce Back as NATO Threats Drive Army Out," *Washington Post*, November 18, 1998; and Judah, *Kosovo*, 189.

140. Guy Dinmore, "Villagers Slaughtered in Kosovo 'Atrocity'; Scores Dead in Bloodiest Spree of Conflict," *The Washington Post*, January 17, 1999.

141. Rexhep Selimi interview, May 18, 2010.

142. OSCE Kosovo Verification Mission, *Kosovo as Seen as Told*, chapters 1, 7.

143. Gellman Barton, "U.S. Has 'Vital Interests' in Containing Conflict," *Washington Post*, February 21, 1999.

144. Madeleine Albright, *Madam Secretary* (New York: Miramax Books, 2003), 502.

145. State Department, *Interim Agreement for Peace and Self-Government in Kosovo*, February 29, 1999; retrieved on December 22, 2014, from http://peacemaker.un.org/kosovo-rambouilletagreement99.

146. Judah, *Kosovo*, 152; and Stephen Hosmer, *The Conflict over Kosovo: Why Milosevic Decided to Settle When he Did* (Santa Monica, CA: RAND, 2001), 11.

147. Slavoljub Djukic, *Milosevic and Markovic* (Montreal: McGill-Queen's University Press, 2001), 127; Jane Perlez, "Purges Hint at Beginning of the End for Milosevic," *New York Times*, November 29, 1998; and Cohen, *Serpent in the Bosom*, 250–257.

148. John Diamond, "Yugoslavia, Iraq Talked Air Defense Strategy," *Philadelphia Inquirer*, March 30, 1999.

149. William Clinton, "Address to the Nation on Airstrikes against Serbian Targets in the Federal Republic of Yugoslavia (Serbia and Montenegro)," *The American Presidency Project*, March 24, 1999.

150. Albright titled the chapter in her memoirs concerning Kosovo "Milosevic Is the Problem."

151. U.S. State Department, *Rambouillet Agreement*, Chapter 8 Article I. 3.

152. Albright, *Madam Secretary*, 513.

153. Rexhep Selimi interview, May 18, 2010.

154. Headquarters U.S. Air Force, *The Air War over Serbia: Aerospace Power in Operation Allied Force: Initial Report* (Ramstein, Germany: USAF Europe Studies and Analysis Directorate, 2000), 9; Paul Strickland, "USAF Aerospace–Power Doctrine: Decisive or Coercive?" *Aerospace Power Journal* 14 (2000): 16, 21; and Benjamin Lambeth, *NATO's Air War for Kosovo* (Santa Monica, CA: RAND, 2001), 21.

155. Headquarters U.S. Air Force, *The Air War over Serbia*, 17; and Robert Hewson, "Operation Allied Force: The First 30 Days," *World Air Power Journal* 38 (1999): 16.

156. Lambeth, *NATO'S Air War for Kosovo*, 17.

157. Headquarters U.S. Air Force, *The Air War over Serbia*, 11.

158. Lambeth, *NATO'S Air War for Kosovo*, 22.

159. "Kosovo Update," *New York Times*, April 9, 1999.

160. Steven Erlanger, " Support for Homeland Up as Sirens Wail and News Is Censored," *New York Times*, March 29, 1999.

161. Jeffrey Smith and William Drozdiak, "Serbs' Offensive Was Meticulously Planned," *Washington Post*, April 11, 1999.

162. Frank Scott Douglas, *Hitting Home: Coercive Theory, Air Power, and Authoritarian Targets*. Dissertation (New York: Columbia University, 2006), 487; "Kosovo Update," *New York Times*, April 7, 1999; Judah, *Kosovo: War and Revenge*, 240; and Office

of U.S. Foreign Disaster Assistance, "Kosovo Crisis Fact Sheet #66," June 11, 1999. By the war's end in June, an estimated 1.3 million Kosovars had fled their homes.

163. Christopher Haave and Phil Haun, eds., *A-10s over Kosovo* (Montgomery AFB, AL: Air University Press, 2003).

164. Ivo Daalder and Michael O'Hanlon, *Winning Ugly: NATO's War to Save Kosovo* (Washington, DC: Brookings Institute, 2000), 136.

165. William Arkin, "Operation Allied Force: 'The Most Precise Application of Air Power in History.'" In *War Over Kosovo: Politics and Strategy in a Global Age,* edited by Andrew Bacevich and Eliot Cohen (New York: Columbia University Press, 2001), 12.

166. Douglas, *Hitting Home,* 495.

167. Bradley Graham, "Missiles Hit State TV, Residence of Milosevic," *Washington Post,* April 23, 1999.

168. "Kosovo Update," *New York Times,* April 26, 1999; and Wesley Clark, *Waging Modern War* (New York: Public Affairs, 2001), 287.

169. Talbott, *The Russia Hand,* 310; and Albright, *Madam Secretary,* 530.

170. Daalder and O'Hanlon, *Winning Ugly,* 140–141. Even so, Clinton still refused to agree to a ground invasion even under pressure from British Prime Minister Tony Blair. Tony Blair, "A Military Alliance and More," *New York Times,* April 24, 1999.

171. William Drozdiak, "Milosevic Foes Urge U.S. to End Sanctions; Measures Said to Help Yugoslav President," *Washington Post,* October 26, 1999. Serbia, however, was in part able to circumvent the most draconian measure, the oil ban, by means of Ukraine shipments along the Danube. Douglas, *Hitting Home,* 530.

172. The number of combat aircraft increased from 300 to over 800. Douglas, *Hitting Home,* 512; and "Kosovo Update," *New York Times,* April 11 and May 4, 1999.

173. John Gordon, Bruce Nardulli, and Walter Perry, "The Operational Challenges of Task Force Hawk," *Joint Forces Quarterly* (Autumn/Winter 2001–2002): 52–57.

174. "Kosovo Update," *New York Times,* May 6 and 16, 1999.

175. Following the Chinese embassy bombing, attacks on Serbian leadership targets practically ceased, the exception being the May 25 and 26 bombings of the Dobanovci presidential villa command and control bunker. However, attacks on Serbian infrastructure continued with multiple strikes on petroleum facilities, electrical power stations, bridges, railways, dual-use factories, and TV and radio stations. NATO HQ, *Operational Updates,* May 8–June 3, 1999; and "Kosovo Update," *New York Times,* May 9 and 12, 1999.

176. Headquarters U.S. Air Force, *The Air War over Serbia.*

177. The USAF employed soft attacks on the grid by dropping CBU-102/B cluster bombs containing BLU-114B submunitions containing aluminum-coated glass fibers, which shorted out, but did not destroy, electrical power stations. Arkin, "Operation Allied Force," 18.

178. International Monetary Fund, *World Economic Outlook Dataset*, April 2009; retrieved on December 22, 2014, from www.imf.org/external/pubs/ft/weo/2009/01/ weodata/index.aspx.

179. Robert Block, "In Belgrade, Hardship Mounts under Air Siege," *Wall Street Journal*, May 12, 1999; and Hosmer, *The Conflict over Kosovo*, 70.

180. Robert Block, "In Belgrade, Hardship Mounts under Air Siege," *Wall Street Journal*, May 12, 1999; and Louis Sell, *Slobodan Milosevic and the Destruction of Yugoslavia* (Durham, NC: Duke University Press, 2002), 311.

181. Carlotta Gall, "Crisis in the Balkans: Serbia; Women Protest Draftees' Kosovo Duty," *New York Times*, May 20, 1999; Carlota Gall, "Crisis in the Balkans: Serbia; Wives Protest and General Sends Troops Back Home," *New York Times*, May 21, 1999; Carlotta Gall, "Crisis in the Balkans: Serbia; Protests Are Resumed by Families of Reservists Ordered Back to Duty in Kosovo," *New York Times*, May 25, 1999; and Blaine Harden, "Reservists a Crucial Factor in Effort against Milosevic," and "Trouble in the Backyard," *New York Times*, July 8 and 9, 1999.

182. Steven Erlanger, "Milosevic Abruptly Fires a High-Profile Maverick," *New York Times*, April 29, 1999.

183. Robert Block, "Serb Official Urges Deal on Kosovo Peace Force—Close Milosevic Associate Backs UN Contingent; U.S. Troops a Possibility," *Wall Street Journal*, May 14, 1999.

184. Steven Erlanger, "Yugoslav Politicians Carefully Maneuver for Day Milosevic Is Gone," *New York Times*, May 21, 1999.

185. "Kosovo Update," *New York Times*, April 15, 1999.

186. Talbott, *The Russia Hand*, 314; and Albright, *Madam Secretary*, 530.

187. The G8 included the United States, Britain, Germany, France, Japan, Italy, Canada, and Russia. G8 Foreign Ministers, *Statement by the Chairman on the Conclusion of the Meeting of the G8 Foreign Ministers on the Petersberg*, May 6, 1999; and Roger Cohen, "Allies and Russia Planning Statement on Kosovo Force," *New York Times*, May 6, 1999.

188. G8 Foreign Ministers, *Statement by the Chairman on the Conclusion of the Meeting of the G8 Foreign Ministers on the Petersberg*, May 6, 1999.

189. "Kosovo Update," *New York Times*, May 20, 1999.

190. John Broder and Jane Perlez, "In Washington, Wary Reaction but Also Relief," *New York Times*, June 4, 1999.

191. Clark, *Waging Modern War*, 358–370.

192. UN Security Council, "Resolution 1244 (10 June 1999)."

193. Clark, *Waging Modern War*, 373.

194. Daniel Byman and Matthew Waxman, "Kosovo and the Great Air Power Debate," *International Security* 24 (2001); Anthony Cordesman, *The Lessons and Non-Lessons of the Air and Missile Campaign in Kosovo* (Westport, CT: Praeger, 2001);

Daalder and O'Hanlon, *Winning Ugly*; Lambeth, *NATO'S Air War for Kosovo*; Andrew Stigler, "A Clear Victory for Air Power: NATO's Empty Threat to Invade Kosovo," *International Security* 27 (2002/2003); Barry Posen, "The War for Kosovo: Serbia's Political-Military Strategy," *International Security* 24 (2000); James Kurth, "First War of the Global Era: Kosovo and U.S. Grand Strategy." In *War Over Kosovo*, edited by Andrew Bacevich and Eliot Cohen (New York: Columbia University Press, 2001); Hosmer, *The Conflict over Kosovo*; Douglas, *Hitting Home*; and Daniel Lake, "The Limits of Coercive Airpower: NATO'S "Victory" in Kosovo Revisited," *International Security* 34 (2009).

195. The author flew combat missions in support of the KLA offensive and witnessed the inability of the KLA to penetrate more than a few miles past the Albanian–Kosovo border.

196. Kelly Greenhill, "The Use of Refugees as Political and Military Weapons in the Kosovo Conflict." In *Yugoslavia Unraveled: Sovereignty, Self-Determination, Intervention*, edited by Raju Thomas (Lanham, MD: Lexington Books, 2003), 215.

197. Steven Lee Myers, "NATO Commander Says Train Was Hit Not Once, but Twice," *New York Times*, April 13, 1999.

198. Michael Gordon, "Civilians Are Slain in Military Attack on a Kosovo Road," *New York Times*, April 14, 1999; and Michael Gordon, "NATO Admits It Hit 2d Convoy in Kosovo," *New York Times*, April 20, 1999.

199. Steven Erlanger, "Blackened Bodies and a Half-Eaten Meal," *New York Times*, April 16, 1999.

200. A further example is NATO bombs hitting so close as to shatter windows of the Swiss and Swedish ambassadors' residences. "Kosovo Update," *New York Times*, May 21, 1999.

201. Michael Gordon, "Russian Anger at U.S. Tempered by Need for Cash," *New York Times*, March 25, 1999; John Broder, "A Phone Call from Gore and a U-Turn to Moscow," *New York Times*, March 24, 1999.

202. Celestine Bohlen, "Yeltsin Sends His Premier to Urge Serbs to Negotiate," *New York Times*, March 29, 1999.

203. Celestine Bohlen, "'Don't Push Us,' Yeltsin Warns West on Balkans," *New York Times*, April 10, 1999.

204. Bennett, *Yugoslavia's Bloody Collapse*, 133.

205. NATO targeted factories owned by Milosevic's wealthy political supporters, even calling and faxing factory owners to be aware of NATO's intentions. There is not, however, evidence that Milosevic conceded as a result of crony attacks. For arguments for crony attack, see Hosmer, *The Conflict over Kosovo*, "Chapter 6: Damage to 'Dual-Use' Infrastructure Generated Growing Pressure"; and Douglas, *Hitting Home*.

206. Robert Pape, "The True Worth of Air Power," *Foreign Affairs* 83 (2004): 2.

207. Katharine Seelye, "Clinton Keeps Option for Ground Troops," *New York Times*, May 19, 1999.

208. Jane Perlez, "Clinton Is Pushing for 50,000 Troops at Kosovo Border," *New York Times*, May 21, 1999.

209. Jane Perlez, "Clinton and the Joint Chiefs to Discuss Ground Invasion," *New York Times*, June 1, 1999; and Clark, *Waging Modern War*, 310.

210. General Wesley Clark has alternatively argued that the KLA operations may well have convinced Milosevic of a ground invasion. Dana Priest, "Kosovo Land Threat May Have Won War," *Washington Post*, September 19, 1999.

211. Slobodan Milosevic, "Yugoslav President Slobodan Milosevic's Address to the Nation," *Washington Post*, June 10, 1999.

Chapter 6

1. Charles Hermann, *Crises in Foreign Policy: A Simulation Analysis* (Indianapolis: Bobbs-Merrill, 1969), 414.

2. Dirk Vandewalle, ed., *Libya since 1969: Qadhafi's Revolution Revisited* (New York: Palgrave MacMillan, 2008), 35.

3. Muammar Qaddafi, *The Green Book* (London: Martin, Brian & O'Keeffe, 1976), 33.

4. Henry Tanner, "Libyan Predicts Oil Will Become Defense Weapon: Qaddafi, at a Long Meeting with Press, Cites Right to Nationalize Resources," *New York Times*, May 14, 1973; and "Bunker Hill Nationalization Will Cause $4-Billion Loss," *New York Times*, June 15, 1973.

5. Vandewalle, *Libya since Independence: Oil and State Building* (Ithaca, NY: Cornell University Press, 1998), 66.

6. Ibid., 35.

7. Ibid.; and Edward Haley, *Qaddafi and the United States since 1969* (New York: Praeger, 1984), 224.

8. "U.S. Diplomats Leave Libya," *New York Times*, February 8, 1980.

9. Bob Woodward, *Veil: The Secret Wars of the CIA 1981–1987* (London: Headline, 1987), 96.

10. Brian L. Davis, *Qaddafi, Terrorism, and the Origins of the U.S. Attack on Libya* (New York: Praeger, 1990), 39.

11. Aftab Kamal Pasha, *Libya and the United States: Qadhafi's Response to Reagan's Challenge* (New Delhi: Détente Publications, 1984), 8.

12. Warren Weaver, "International Dispute Is Centered on Status of Mediterranean Gulf," *New York Times*, August 19, 1981.

13. Jack Anderson, "Qaddafi Is Said to Voice Threat against Reagan," *Washington Post*, October 13, 1981; and Woodward, *Veil*, 167.

14. Bernard Gwertzman, "U.S. Decision to Embargo Libyan Oil Is Reported; Embargo Decision Reported," *New York Times*, February 26, 1982.

15. Tim Niblock, *"Pariah States" & Sanctions in the Middle East: Iraq, Libya, Sudan* (Boulder, CO: Lynne Rienner, 2001), 28–29.

16. Bernard Gwertzman, "Brazil Grounds 4 Libyan Planes Carrying Arms; U.S. Had Sought Ban on Flights to Nicaragua Brazil to Block Libya Arms Flight," *New York Times*, April 21, 1983; and Haley, *Qaddafi and the United States since 1969*, 319–321.

17. Charles Mohrs, "Marines' Security Raises Questions," *New York Times*, October 24, 1984.

18. Ronald Reagan, "National Security Council—National Security Decision Directive on Combating Terrorism," April 26, 1984.

19. Jon Nordheimer, "Gunman in London in Libyan Embassy Fires into Crowd: A Police Officer Is Killed," *New York Times*, April 18, 1984.

20. John Tagliabue, "Ex-Captives Say Gunmen Planned to Kill Military Men One by One," *New York Times*, July 2, 1985; John Tagliabue, "Appeal by Captain: Unconfirmed Reports Say Some on Board May Have Been Slain," *New York Times*, October 8, 1985; Associated Press, "Egyptian Jet Hijacked to Malta; 3 or 4 Aboard Are Reported Slain," *New York Times*, November 24, 1985; and Wolfgang Saxon, "Airport Attacks Widely Deplored," *New York Times*, December 28, 1985.

21. Ronald Reagan, "The President's News Conference," January 7, 1986.

22. National Security Council, "National Security Decision Direction 205: Annex: Acting against Libyan Support of International Terrorism," January 8, 1986.

23. Bernard Weinraub, "In Disputed Area: Libya Says It Downed 3 Jets, but Washington Reports No Losses," *New York Times*, March 25, 1986.

24. Woodward, *Veil*, 444.

25. Joseph Stanik, *El Dorado Canyon: Reagan's Undeclared War with Qaddafi* (Annapolis, MD: Naval Institute Press, 2003), 150.

26. Ibid., 183.

27. Ibid., chapter 6.

28. Foreign Broadcast Information Service (FBIS), "Al Qadhdhafi Appears on TV," *Daily Report* FBIS-SOV-86-074, April 17, 1986.

29. FBIS, "Civilians, Embassies Targeted," *Daily Report* FBIS-SAS-86-072, April 15, 1986; Stanik, *El Dorado Canyon*, 189, 207; and Davis, *Qaddafi*, 141.

30. Davis, *Qaddafi*, 139.

31. Daniel Bolger, *Americans at War: 1975–1986, An Era of Violent Peace* (Novato, CA: Presidio Press, 1988), 423; Stanik, *El Dorado Canyon*, 192-193; and Davis, *Qaddafi*, 140.

32. Judith Miller, "Italian Island, a Libyan Target, Escapes Unscathed," *New York Times*, April 16, 1986; and E. J. Dionne, "Italian Promises to Answer Terror," *New York Times*, April 20, 1986.

33. FBIS—Middle East & Africa, "Foreign Liaison Bureau Issues Statement on Raids," and "Attack Everything American," *Daily Report* FBIS-MEA-86-072, April 15, 1986.

34. BBC, "1986: British Journalist McCarthy Kidnapped," BBC online, April 17, 1986; and Davis, *Qaddafi*, 158.

35. U.S. Office of Secretary of State, *Patterns of Global Terrorism: 1987* (Washington, DC: Author, 1988), 6; and U.S. Office of Secretary of State, *Patterns of Global Terrorism: 1988*, 44.

36. U.S. Office of Secretary of State, "Introduction," in *Patterns of Global Terrorism: 1987*; and U.S. Office of Secretary of State, "Introduction," in *Patterns of Global Terrorism: 1988*.

37. U.S. Office of Secretary of State, *Patterns of Global Terrorism: 1988*, 44.

38. Bruce Jentleson, "The Reagan Administration and Coercive Diplomacy: Restraining More than Remaking Governments," *Political Science Quarterly* 106 (1991): 64.

39. FBIS, "Fighting Reported in Tripoli between Factions," *Daily Report* FBIS-MEA-86-073, April 16, 1986.

40. Bob Woodward, "Gadhafi Target of Secret U.S. Deception Plan," *Washington Post*, October 2, 1986.

41. Woodward, *Veil*, 474–475; and Ronald Reagan, "Libya Policy," National Security Decision Directives 234, August 16, 1986. The content of NSD Directive 234 remains classified.

42. Woodward, "Gadhafi Target of Secret U.S. Deception Plan."

43. Barbara Geddes, "Authoritarian Breakdown: Empirical Test of a Game Theoretic Argument." Paper presented at the Annual Meeting of the American Political Science Association, Atlanta, GA, April 1999.

44. Jessica Weeks, "Autocratic Audience Costs: Regime Type and Signaling Resolve," *International Organization* 62 (2008): 35–64.

45. CNN, "U.S. Policy on Assassinations," CNN.COM Law Center, November 4, 2002.

46. Niblock, *"Pariah States,"* 24.

47. David Horovitz, "Gaddafi Personally Okayed Lockerbie Bombing," *The Jerusalem Post*, September 4, 2009.

48. U.S. Representative to the United Nations, "United States District Court for the District of Columbia Indictment for Abdel Basset and Lamen Fhimah," UN General Assembly Security Council A/46/831 S/23317, December 23, 1991.

49. Office of Press Secretary, "Statement Announcing Joint Declarations on the Libyan Indictments," American Presidency Project, November 27, 1991.

50. "Qaddafi Scoffs at Demands for Bombing Suspects," *New York Times*, November 29, 1991.

51. Tripoli JANA (Jamahaniyyah News Agency), "People's Bureau Denies Locker-bie Involvement," Daily Reports FBIS-NES-91-221, November 15, 1991.

52. Associated Press, "Libya Denies Involvement in Pan Am Bombing," *Washington Post*, November 16, 1991.

53. Niblock, *"Pariah States,"* 37.

54. Hans Koechler, "Memorandum dated 6 February 1992 from the President of the International Progress Organization," UN Security Council A/46/886 S/23641, February 6, 1992.

55. UN Security Council, "Resolution 731 (1992)," January 21, 1992.

56. Boutros Boutros-Ghali (February 11, 1992), "Report by the Secretary-General Pursuant to Paragraph 4 of Security Council Resolution 731 (1992)," UN Security Council S/23574, February 11, 1992.

57. Niblock, *"Pariah States,"* 39.

58. UN Security Council, "Resolution 748 (1992)."

59. Niblock, *"Pariah States,"* 41.

60. UN Security Council (November 11, 1993), "Resolution 883 (1993)."

61. West Texas Research Group, "Oil Price History and Analysis 2008," WTRG Economics; Niblock, *"Pariah States,"* 65.

62. International Monetary Fund, "Report for Libya," *World Economic Outlook Database*, October 2009.

63. Ray Takey, "Qadhafi and the Challenge of Militant Islam," *Washington Quarterly* 21 (1998): 164; and Wyn Bowen, "Libya & Nuclear Proliferation: Stepping Back from the Brink," *Adelphi Papers* 46 (2006), 55.

64. Takey, "Qadhafi and the Challenge of Militant Islam," 168.

65. Bruce Jentleson and Christopher Whytock, "Who 'Won' Libya? The Force–Diplomacy Debate and Its Implications for Theory and Policy," *International Security* 30 (2005/2006): 66.

66. Once the bill passed the House of Representatives, President Clinton signed it into law on August 5, 1996. Kenneth Katzman, "The Iran–Libya Sanctions Act (ILSA)," *CRS Report for Congress* Order Code RS20871, April 26, 2006.

67. Charles Doyle, "Antiterrorism and Effective Death Penalty Act of 1996: A Summary," Congressional Research Service, June 3, 1996.

68. Gary Hufbauer, Jeffrey Schott, Kimberly Elliott, and Barbara Oegg, "Case 78-8 United States v. Libya" and "Case 92-12 United Nations v. Libya," in *Economic Sanctions Reconsidered*, 3rd ed. (Washington, DC: Peterson Institute, 2008).

69. Douglas Jehl, "Arab Countries Vote to Defy U.N. Sanctions against Libya," *New York Times*, September 22, 1997.

70. "World Court Claims Jurisdiction in Pan Am Flight 103 Bombing," *New York Times*, February 28, 1998.

71. UN Security Council, "Resolution 1192 (1998)" S/RES/1192, August 27, 1998.

72. Niblock, *"Pariah States,"* 51.

73. For details on the negotiations, see Hufbauer, "Case 78-8 United States v. Libya" and "Case 92-12 United Nations v. Libya," in *Economic Sanctions Reconsidered.*

74. Judith Miller, "In Rare Talks with Libyans, U.S. Airs Views on Sanctions," *New York Times,* June 12, 1999.

75. Gordon Corera, *Shopping for Bombs: Nuclear Proliferation, Global Insecurity, and the Rise and Fall of the A. Q. Khan Network* (Oxford, UK: Oxford University Press, 2006), 180.

76. Yahia Zoubir, "The United States and Libya: From Confrontation to Normalization," *Middle East Policy* vol. XIII (2006): 50; and IAEA Board of Governors, "Implementation of the NPF Safeguards Agreement of the Socialist People's Libyan Aram Jamahiriya," GOV/2004/12, February 20, 2004, 5.

77. Ronald St John, "'Libya Is Not Iraq': Preemptive Strikes, WMD and Diplomacy," *The Middle East Journal* 58 (2004): 399.

78. Ronald Neumann, "Libya: A U.S. Policy Perspective," *Middle East Policy* 7 (2000): 143–145.

79. Barbara Slavin, "Libya's Rehabilitation in the Works since Early '90s," *USA Today,* April 26, 2004.

80. Bowen, "Libya & Nuclear Proliferation," 8.

81. Maalfried Braut-Hegghamer, "Libya's Nuclear Intentions: Ambition and Ambivalence," *Strategic Insights* vol. VIII (2009): 62.

82. Ronald St John, "The Soviet Penetration of Libya," *The World Today* 38 (1982): 135; and IAEA, "Libyan Arab Jamahiriya, Socialist People's: IRT-1," *Nuclear Research Reactors in the World,* 1999.

83. Dirk Vandewalle, ed., *Libya since 1969: Qadhafi's Revolution Revisited* (New York: Palgrave MacMillan, 2008), 35.

84. Bowen, "Libya & Nuclear Proliferation," 31–32.

85. FBIS Daily Report, "Al-Qadhdhafi Wants Long-Range Arab 'Missile'" *Tripoli Television Service* FBIS-NES-90-078, April 23, 1990.

86. William Cohen, "Proliferation: Threat and Response 1997," *Secretary of Defense* (Washington, DC: U.S. Government Publications, 1997).

87. Nuclear Threat Initiative, "Libya Profile Chemical Overview," *NTI Country Profiles,* September 2009.

88. Nuclear Threat Initiative, "Libya Profile Biological Overview."

89. Corera, "Chapter 8: Dealing with Gadaffi," in *Shopping for Bombs.*

90. Bowen, "Libya & Nuclear Proliferation," 37.

91. Ibid., 44.

92. Nuclear Threat Initiative, "Libya Profile Chemical Overview."

93. Leonard Spector and Jacqueline Smith, *Nuclear Ambitions: The Spread of Nuclear Weapons 1989–1990* (Boulder, CO: Westview, 1990), 179.

94. William S. Perry, "Proliferation: Threat and Response," *Secretary of Defense* (Washington, DC: U.S. Government Publications, 1996). Libya's disclosures to the IAEA in 2004 showed no evidence of any biological weapons program.

95. Slavin, "Libya's Rehabilitation in the Works since Early '90s."

96. Yahia Zoubir, "The United States and Libya: From Confrontation to Normalization," 57.

97. Slavin, "Libya's Rehabilitation in the Works since Early '90s."

98. Zoubir, "The United States and Libya: From Confrontation to Normalization," 59.

99. Stephen Fidler, Mark Huband, and Roula Khalaf, "Return to the Fold: How Gadaffi Was Persuaded to Give up His Nuclear Goals," *Financial Times*, January 27, 2004.

100. Bowen, "Libya & Nuclear Proliferation," 62.

101. Michael Hirsch, "Bolton's British Problem," *Newsweek*, 145 (2005), 30.

102. White House, "Statement by the Press Secretary," August 15, 2003.

103. UN Security Council, "Resolution 1506 (2003)," September 12, 2003.

104. "Libya Gives Up Chemical Weapons," *BBC News*, December 19, 2003.

105. George W. Bush, "Transcript: First Presidential Debate," *Washington Post*, September 30, 2004; and Richard Cheney, "Transcript: Vice Presidential Debate," *Washington Post*, October 5, 2004.

106. St John, "'Libya Is Not Iraq,'" 386–402.

107. George Joffe, "Libya: Who Blinked, and Why," *Current History* 103 (2004): 221–225; and St John, "'Libya Is Not Iraq,'" 386–402.

108. Jentleson, "Who 'Won' Libya?" 47–86.

109. Brit Hume, "Muammar Qaddafi: 'I Saw Iraq and I Was Afraid,'" *Fox News*, December 26, 2003.

110. Martin Indyk, "The Iraq War Did Not Force Gadaffi's Hand," *Financial Times*, March 9, 2004.

111. Ibid.

112. St John, "Libya Is Not Iraq," 387.

113. Bowen, "Libya & Nuclear Proliferation," 67.

114. Jentleson, "Who 'Won' Libya?" 47–86.

115. Michael Hirsch, "Bolton's British Problem," 30.

116. U.S. Government Accountability Office, "Figure 3: Average Number of Daily, Enemy-Initiated Attacks against the Coalition, Iraq Security Forces, and Civilians," in *Securing, Stabilizing, and Rebuilding Iraq* GAO-07-1195 (September 2007), 11.

Chapter 7

1. Geoffrey Blainey, *The Causes of War* (New York: Free Press, 1988), 56.

2. UN Security Council, "Resolution 678 (1990), 29 November 1990"; George Bush and Brent Scowcroft, *A World Transformed* (Alfred A. Knopf: New York, 1998), 394; and Bob Woodward, *The Commanders* (Simon and Schuster: New York, 1991), 320.

3. Kevin Woods and Mark Stout, "Saddam's Perceptions and Misperceptions: The Case of 'Desert Storm.'" *Journal of Strategic Studies* 33 (2010): 12.

4. Daryl Press, *Calculating Credibility: How Leaders Assess Military Threats* (Ithaca, NY: Cornell University Press, 2005), 2.

5. Barack Obama, "Transcript: President Obama's Address to the Nation on Military Action in Libya," *ABC Nightline*, March 28, 2011.

6. UN Security Council, "Resolution 1970 (2011), 26 February 2011"; and UN Security Council, "Resolution 1973 (2011), 17 March 2011."

Appendix A

1. 1950 Korean War, 1961 Bay of Pigs, 1979 Iranian Hostage Crisis, 1983 Grenada Invasion,1989 Panamanian Invasion, 1992 Iraq No-Fly Zone, and 1996 Iraqi Desert Strike.

2. Michael Brecher and Jonathan Wilkenfeld, *A Study of Crisis* (Ann Arbor: University of Michigan Press, 1997, 2000); available online at the International Crisis Behavior (ICB) Project www.cidcm.umd.edu/icb/, ICB # 210; George C. Herring, *America's Longest War: The United States and Vietnam, 1950–1975* (New York: McGraw-Hill, 1996), 126–141; Robert D. Schulzinger, *A Time for War: The United States and Vietnam, 1941–1975* (New York: Oxford University Press, 1997), 142–170; and Mark Clodfelter, *The Limits of Air Power: The American Bombing of North Vietnam* (New York: The Free Press, 1989), 45–72.

3. Clodfelter, *The Limits of Air Power*, 48.

4. Herring, *America's Longest War*, 133–138.

5. Ibid., 138.

6. ICB # 213; Herring, *America's Longest War*; Schulzinger, *A Time for War*; Clodfelter, *The Limits of Air Power*; Robert A. Pape, *Bombing to Win: Air Power and Coercion in War* (Ithaca, NY: Cornell University Press, 1996); Lien-Hang Nguyen, *Hanoi's War* (Chapel Hill: University of North Carolina Press, 2012); Earl H. Tilford, *Setup: What the Air Force Did in Vietnam and Why* (Montgomery, AL: Air University Press, 1991); and The Official History of the People's Army of Vietnam, 1954–1975, *Victory in Vietnam: The Official History of the People's Army of Vietnam, 1954–1975* (Lawrence: University Press of Kansas, 2002).

7. Clodfelter, *The Limits of Air Power*, 65.

8. Ibid., 88–115.

9. ICB #225; Herring, *America's Longest War*; Schulzinger, *A Time for War*; and Clodfelter, *The Limits of Air Power*.

10. ICB# 238; Matthew Bonham and Michael Shapiro, "Explanations of the Unexpected: The Syrian Intervention in Jordan 1970," in *Structure of Decision: The Cognitive Maps of Political Elites*, edited by Robert Axelrod (Princeton, NJ: Princeton University Press, 1976); Henry Brandon, *The Retreat of American Power* (New York: Doubleday, 1972); Henry Brandon and David Schoenbaum, "Jordan: The Forgotten Crisis: I," *Foreign Policy* X (1973); Alan Dowty, *Middle East Crisis: US Decision-Making in 1958, 1970, and 1973* (Berkeley, CA: University of California Press, 1984); Adam M. Garfinkle, "US Decision Making in the Jordan Crisis: Correcting the Record," *Political Science Quarterly* 100:1 (Spring 1985): 171–178; Paul K. Huth, *Extended Deterrence and the Prevention of War* (New Haven, CT: Yale University Press, 1988); Walter Isaacson, *Kissinger: A Biography* (New York: Simon & Schuster, 1992); Bernard Kalb and Marvin Kalb, *Kissinger* (Boston: Little Brown & Co., 1974); Henry Kissinger, *White House Years* (New York: Simon & Schuster, 1979); Douglas Little, "A Puppet in Search of a Puppeteer? The United States, King Hussein, and Jordan, 1953–1970," *The International History Review* 17:3 (August 1995); and Tad Szulc, *The Illusion of Peace: Foreign Policy in the Nixon Years* (New York: Viking Press, 1978).

11. Kissinger, *White House Years*, 601.

12. Henry Brandon, "Jordan: The Forgotten Crisis: Were We Masterful . . ." *Foreign Policy* 10 (1973): 159.

13. Adam M. Garfinkle, "US Decision Making in the Jordan Crisis: Correcting the Record," *Political Science Quarterly* 100:1 (1985): 124.

14. Brandon, "Jordan: The Forgotten Crisis," 167.

15. ICB #246; Herring, *America's Longest War*; Schulzinger, *A Time for War*; Clodfelter, *Limits of Air Power*; Pape, *Bombing to Win*; Kissinger, *The White House Years*; Tilford, *Setup*; The Official History of the People's Army of Vietnam; and Stephen Randolph, *Powerful and Brutal Weapons: Nixon, Kissinger, and the Easter Offensive* (Cambridge, MA: Harvard University Press, 2007).

16. ICB #249; Pierre Asselin, *A Bitter Peace: Washington, Hanoi, and the Making of the Paris Agreement* (Chapel Hill: University of North Carolina Press, 2003); Herring, *America's Longest War*; Schulzinger, *A Time for War*; Clodfelter, *The Limits of Air Power*; Pape, *Bombing to Win*; Kissinger, *The White House Years*; Nguyen, *Hanoi's War*; Tilford, *Setup*; The Official History of the People's Army of Vietnam, 1954–1975; and Marshall Michel, *The 11 Days of Christmas: America's Last Vietnam Battle* (San Francisco: Encounter Books, 2002).

17. ICB# 259; U.S. Congress, *Warpowers: A Test of Compliance Relative to the Danang Sealift, the Evacuation of Phnom Penh, the Evacuation of Saigon, and the Mayaguez Incident*, Hearings, 94th Congress, 1st Session, 1975 (Washington, DC: U.S. Government Printing Office, 1975); Richard G. Head, Frisco Short, and Robert McFarlane,

Crisis Resolution: Presidential Decision Making in the Mayaguez and Korean Confrontations (Boulder, CO: Westview Press, 1978); and Robert R. Simmons, *The Pueblo, EC-121, and Mayaguez Incidents: Some Continuities and Changes* (Baltimore, MD: Maryland International Law Society, 1978).

18. Simmons, *The Pueblo, EC-121, and Mayaguez Incidents*.

19. "White House Says Cambodia Seized a U.S. Cargo Ship," *New York Times*, May 13, 1975, A1.

20. Three Marines were left behind and later killed by the Khmer Rouge.

21. Head, *Crisis Resolution*, 149–215; Richard A. Mobley, "Revisiting the Korean Tree-Trimming Incident," *Joint Forces Quarterly* 35 (2003): 108–115; and John K. Singlaub, *Hazardous Duty: An American Soldier in the Twentieth Century* (New York: Summit, 1991).

22. Head, *Crisis Resolution*, 159–170.

23. Ibid., 198.

24. ICB #330.

25. Bob Woodward, *Veil: The Secret Wars of the CIA 1981–1987* (London: Headline, 1987), 96.

26. Brian L. Davis, *Qaddafi, Terrorism, and the Origins of the U.S. Attack on Libya* (New York: Praeger, 1990), 39.

27. Aftab Kamal Pasha, *Libya and the United States: Qaddafi's Response to Reagan's Challenge* (New Delhi: Détente Publications, 1984), 8.

28. Warren Weaver, "International Dispute Is Centered on Status of Mediterranean Gulf," *New York Times*, August 19, 1981.

29. Jack Anderson, "Qaddafi Is Said to Voice Threat against Reagan," *Washington Post*, October 13, 1981; and Woodward, *Veil*, 167.

30. ICB # 383, 384; Bernard Gwertzman, "U.S. Decision to Embargo Libyan Oil Is Reported; Embargo Decision Reported," *New York Times*, February 26, 1982.

31. Kenneth Roberts, "Bullying and Bargaining: The United States, Nicaragua, and Conflict Resolution in Central America," *International Security* 15 (2) Autumn 1990: 67–102; Johanna Oliver, "The Esquipulas Process: A Central American Paradigm for Resolving Regional Conflict," *Ethnic Studies Report* 17 (2) July 1999; R. Pardo-Maurer, *The Contras, 1980–1989: A Special Kind of Politics* (New York: Praeger, 1990); Robert A. Kagan, *A Twilight Struggle: American Power and Nicaragua 1977–1990* (New York: Free Press, 1996); Robert A. Pastor, *Condemned to Repetition: The United States and Nicaragua* (Princeton, NJ: Princeton University Press, 1987); William I. Robinson, *A Faustian Bargain: U.S. Intervention in the Nicaraguan Elections and American Foreign Policy in the Post-Cold War Era* (Boulder, CO: Westview, 1992); David Ryan, *US–Sandinista Diplomatic Relations: Voice of Intolerance* (New York: St. Martin's Press, 1995); Timothy C. Brown, *Causes of Continuing Conflict in Nicaragua: A View from the Radical Middle* (Stanford, CA: Hoover Institute, 1995); and Woodward, *Veil*.

32. Oliver, "The Esquipulas Process."

33. Pardo-Maurer, *The Contras*, 97.

34. ICB #408; Don Oderdorfer, *The Two Koreas: A Contemporary History* (New York: Basics, 2001); Joel S. Wit, Daniel B. Poneman, and Robert L. Gallucci, *Going Critical: The First North Korean Nuclear Crisis* (Washington, DC: Brookings Institution Press, 2004); Yoichi Funabashi, *The Peninsula Question: A Chronicle of the Second Korean Nuclear Crisis* (Washington, DC: Brookings Institution Press, 2007); and Charles Kartman, Robert Carlin, and Joel Wit, *A History of KEDO 1994–2006* (Stanford, CA: Center for International Security and Cooperation, 2012).

35. 1993 Democratic People's Republic of Korea (North Korea)–United States Joint Statement adopted in New York on June 11, 1993.

36. ICB #411; John R. Ballard, *Upholding Democracy: The United States Military Campaign in Haiti, 1994–1997* (Westport, CT: Praeger, 1998); Madeleine Albright, *Madam Secretary: A Memoir* (London: Macmillan, 2003); Robert Fatton, *Haiti's Predatory Republic: The Unending Transition to Democracy* (Boulder, CO: Lynne Rienner, 2002); Philippe R. Girard, *Clinton in Haiti: The 1994 U.S. Invasion of Haiti* (New York: Palgrave Macmillan, 2004); Peter Hallward, *Damming the Flood: Haiti, Aristide, and the Politics of Containment* (London: Verson, 2007); John F. Harris, *The Survivor: Bill Clinton in the White House* (New York: Random House, 2005); and David Malone, *Decision-Making in the UN Security Council: The Case of Haiti, 1990–1997* (New York: Oxford University Press, 1998).

37. "UN Security Council Resolution 940," July 31, 1994.

38. ICB #412.

39. Youssef Ibrahim, "Baghdad's Burden—A Special Report; Iraq Is Near Economic Ruin but Hussein Appears Secure," *New York Times*, October 25, 1994.

40. Paul Lewis, "U.N.'s Team in Iraq Sees Arms Gains," *New York Times*, July 26, 1994.

41. Baghdad INA, "'Temporary' Changes Made to Ration Card Quotas," FBIS-NES-94-187 on September 27, 1994; and Ibrahim, "Baghdad's Burden."

42. UN Special Commission, *Report of the Secretary-General on the Status of the Implementation of the Special Commission's Plan for the Ongoing Monitoring and Verification of Iraq's Compliance with Relevant Parts of Section C of the Security Council Resolution 687 (1991) S/1994/1138* (United Nations: Security Council Distribution General, 1994); and Barbara Crossette, "Threats in the Gulf: The U.N.; Iraq's Attempt to Have Sanctions Lifted Quickly May Have Backfired," *New York Times*, October 11, 1994.

43. William Clinton, "Letter to Congressional Leaders on Iraq," *The American Presidency Project*, October 27, 1994; retrieved on December 26, 2014, from www.presidency.ucsb.edu/.

44. Baghdad Republic of Iraq Radio Network, "Tariq 'Aziz Addresses U.N. General Assembly," FBIS-NES-94-196, October 11, 1994.

45. Barbara Crossette, "Iraqis to Accept Kuwait's Borders," *New York Times*, November 11, 1994.

46. ICB #422, 429.

47. Michael Knights, *Cradle of Conflict: Iraq and the Birth of the Modern U.S. Military* (Annapolis, MD: Naval Institute Press, 2005), 171–210.

48. "UN Security Council Resolution 1134," October 23, 1997; and "UN Security Council Resolution 1137," November 12, 1997.

49. "The Deal on Iraq: Chronology; Four Weeks of Tension," *New York Times*, November 21, 1997.

50. "UN Security Council Resolution 1154," March 2, 1998.

51. U.S. State Department, "Chronology of Events Leading to the U.S.-led Attack on Iraq," retrieved on January 8, 1999, from www.state.gov/www/regions/nea/iraqchronyr.html.

52. "UN Security Council Resolution 1194," September 9, 1998.

53. UNSCOM, "Report of the VX Expert Meeting," October 23, 1998; available at www.fas.org/news/un/iraq/s/981026vx/index.html.

54. "UN Security Council Resolution 1205," November 5, 1998.

55. ICB #427; National Commission on Terrorist Attacks upon the United States, *The 9/11 Commission Report* (New York: W. W. Norton, 2004), 57–70; William J. Clinton, *My Life* (New York: Knopf, 2004), 796–805; Albright, *Madame Secretary*, 459.

56. ICB #434; National Commission on Terrorist Attacks upon the United States, *The 9/11 Commission Report*; and Bob Woodward, *Bush at War* (New York: Simon & Schuster, 2002).

57. "UN Security Council Resolution 1970," February 26, 2011; and "UN Security Council Resolution 1973," March 17, 2011.

58. Barack Obama, "Transcript: President Obama's Address to the Nation on Military Action in Libya," *ABC Nightline*, March 28, 2011.

INDEX

Note: page numbers followed by t and f refer t tables and figures respectively. Those followed by n refer to notes, with note number.

The authorized representative in the EU for product safety and compliance is:
Mare Nostrum Group
B.V Doelen 72
4831 GR Breda
The Netherlands

www.ingramcontent.com/pod-product-compliance
Lightning Source LLC
Chambersburg PA
CBHW031545260326
41914CB00002B/276